T0272455

An Insect A Day

An Insect A Day

Dominic Couzens and Gail Ashton

BATSFORD

First published in the United Kingdom in 2024 by
Batsford
43 Great Ormond Street
London WC1N 3HZ
An imprint of Batsford Books Ltd

ISBN: 9781849947947

A CIP catalogue record for this book is available from the British Library.

30 29 28 27 26 25 24
10 9 8 7 6 5 4 3 2 1

Reproduction by Rival Colour Ltd, UK
Printed and bound by Toppan Leefung Printing Ltd, China

This book can be ordered direct from the publisher at the website www.batsfordbooks.com, or try your local bookshop.

Contents

Introduction

Thousands of insects depend upon just a single plant species, or a few related species, to survive. The four-barred knapweed gall fly lays its eggs, as its name implies, in the seed heads of knapweeds.

Insects are everywhere, all the time, every day of the year. We might not see them on a daily basis, but they are still somewhere. They could be underground, in the sky or anywhere in between. Insects are permanent inhabitants of every continent on Earth. They still haven't quite figured out the deep oceans, but everywhere else is fair game; insects have filled pretty much every available niche since their ancestors left the seas and made landfall around 480 million years ago. Of course, we are more likely to see them in warmer conditions, especially if we live in more seasonally extreme areas, because most insects are exothermic – requiring warmth and sunshine to function physically and metabolically. In temperate regions, insects will have more defined windows of phenological behaviour; that is, their reproductive and hibernacular cycles are tied to summer and winter respectively. In equatorial and sub-tropical regions, however, where there is little seasonal fluctuation, insects are active most of the year as temperatures and food sources are enduringly more favourable. Go towards the poles and insects spend most of the year in homeostasis, emerging for a period of weeks or even days in the brief windows of weak sunshine and relative warmth. The beauty of these climatic adaptations is that we can see insect activity nearly all the time, whether it be the warm hum of hoverflies in the spring garden, the ladybird snuggling up cosily into a hollowed plant stem in autumn, or the hippity-hop of snow-fleas across frosty moss in midwinter.

Before we commence our year-long odyssey, let's remind ourselves what an insect actually is. Insects are classified in the kingdom Animalia, and within that the phylum Arthropoda, which is characterized by a segmented body, multiple pairs of jointed legs and a sclerotized (hardened) exoskeleton surrounding a soft body. Insects further distinguish themselves into their own class, Insecta, with a three-sectioned body (head, thorax and abdomen), three pairs of jointed legs and two pairs of wings, putting themselves in a different taxonomic group to similar arthropods, such as spiders and isopods. The subject of wings is something of a grey area, because not all insects have wings, and many appear to have just one pair, the other pair having long ago evolved into something else. The forewings of beetles,

The sheer number of species of insects is remarkable. The elephant hawk-moth *Deilephila elpenor* is one of 1,450 species in its family Sphingidae, and one of 160,000 species of the order Lepidoptera, the butterflies and moths.

for example, have been modified into hard, protective covers beneath which they stow their hindwings. Flies' hindwings have gone the other way, reducing down into tiny, clubbed stalks that seem useless but have evolved into super-sensitive sensory organs called halteres, which act as gyroscopes and send real-time flight and atmospheric data straight to the pilot's brain. So, while we can't always see wings in insects, what we can be sure of is that if it's small, has wings and isn't a bird or a bat, it's definitely an insect.

Writing about insects has been, for two entomophiles, the easy part – for who could fail to be interested, intrigued, fascinated or even fixated with this group of animals? The hard part has been choosing which species to include, because the maths involving insects are truly staggering. One in two living creatures on Earth is an insect and they make up 90 per cent of all known animal species. The number of known global insect species currently sits at around 1 million, and it is possible that we are barely scratching the surface. Estimates fluctuate wildly between 2 and 30 million insect species on Earth right now, and that doesn't include the many species that are becoming extinct before we've even discovered them. According to the Royal Entomological Society, there are around 1.4 billion insects per human on Earth; their mass outweighs humans 70 times. Other sources estimate that there could be 10 quintillion individual living insects on Earth as you read this – that's a ten with 18 zeros on the end of it. We still don't even talk about money or data in these terms: humans are lagging way behind with mere billions and trillions as a measure of hugeness. Social wasp nests can house tens of thousands of individuals, and termite mounds, hundreds of thousands. So how can our planet house so many insects when we humans are struggling to balance the ecological scales with a global population of 8 billion?

It comes down to size. We tend to measure success on scale – bigger is better and all that – but the foundations of this theory become very shaky when insects are brought into the picture. Insects used to be a lot bigger; take, for example, the gargantuan Meganisoptera of the late Carboniferous period, when atmospheric oxygen levels exceeded 20 per cent. These giant, dragonfly-like beasts had wingspans of 0.5m (½ft) and would give modern-day birds of prey a run for their money. This proved to be an unsustainable body plan, however, as a series of cataclysmic climate events and decreasing atmospheric oxygen levels wiped out the majority of species on Earth. The largest insects were pushing the limits of a successful body plan anyway (the rules of insect biomechanics can only go so far), so over hundreds of millions of years they became smaller, needing less food, water, free oxygen and space to survive. Combined with their short life span and prolific reproductive rate, which allow for rapid adaptation, insects have processed along a wildly successful avenue of evolution that brings us to the modern day, with an almost unquantifiable global population of insects, the diversity and variety of which is truly staggering.

It is so incredibly sad, then, that instead of admiring the success of the insects of Earth, we choose to demonize them. In the very short span of history that humans have occupied the planet (around 2 million years, a little shorter than the near 400-million-year occupation of insects), we have put our six-legged roommates in Room 101, the sin bin, the bad books. We blame them for everything from itchy skin to global food shortages, and our self-styled solution has been to kill them. It is our absolute conviction they are The Enemy, and thus we spend a disproportionate quantity of our time and resources obliterating them from existence, and this is a huge problem. We swat, spray, splat and squash them without really thinking about the consequences of this wholesale insecticide, because we only see the short-term benefits of having fewer insects bothering us. But learn more about insects, and look closely at them, and the veil of disgust will quickly fall away to be replaced with fascination, respect and yes, even affection. Because insects are beautiful! They are fluffy and have big eyes and they often look at you quizzically – they are basically tiny puppies. And many will sit happily on your hand; they can be very good company. They are superb parents, laying their eggs strategically, cleaning and tending their young and protecting them from harm, just like we do. And relatively, they don't bother us at all. For every delirious, hypoglycaemic wasp that panics in our terrifying presence and stings us, and every fiercely determined female mosquito who desperately needs to catalyze her egg production with a blood meal, there are literally millions of other insects that never come near us and have no desire to do so, they just want to get through the day. Sound familiar?

Most of the species in this book will never cross paths with each other, or even be aware of each other's existence, but they all share a strong bond, because they are all part of an expansive and intricate web of life, in which every individual insect plays a critical role. It may become food for something else; it may inadvertently assist in driving global food production through pollination services; it could be one of the recycling team that processes matter into soil nutrients, before its own body ends up as particles of loamy, nutritious compost. It might be one of a legion of predatory or parasitic insects, which naturally suppress population explosions, and scaffold food security within agricultural systems. It could even be one whose toxic biochemistry is being synthesized into the new generation of cancer-killing drugs. Every one of these roles is minor, but together they have formed the glorious, diverse, stable, habitable planet upon which we live today. I don't know about you, but I think that deserves respect and kindness, rather than a fly swatter or the sole of a shoe.

It is with delight and pride that we present to you 366 insects and stories of insect folklore; one for every day of the Gregorian year – including that quadrennial leap day that offers us an extra 24-hour window and – in this book – a bonus opportunity to meet another incredible insect. Our selection contains six-legged ambassadors from all over the globe, all with astonishing adaptations. You will meet a ghostly butterfly that haunts the permanent twilight in the darkest recesses of the tropical rainforests and encounter a surprisingly resilient moth in the frozen Arctic depths. We also dive into the rich cultural fabric that binds us to insects through history via myth, legend and fossil record. If that hasn't piqued your interest, how about a beetle that can conjure its own water supply from the air in the hottest, driest

deserts, and even an insect that has conquered the surface of the ocean? You will read about residents from your own doorstep that you know very well and be introduced to fascinating new species from far-away places. We want you close this book at the end of the year with a new perspective on insects; an increased sense of empathy towards them, an understanding of their place in the world, or maybe even the resolve to go out and discover these extraordinary beasts in the wild. If we can spread our love of this marvellous group of animals further into the world, then we will be very happy.

The lacewing is just one example of the extraordinary evolutionary paths that insects have taken over the last 480 million years.

DOMINIC COUZENS and GAIL ASHTON, 2024

THE FLY

A negative image of a housefly.

It isn't easy for us humans to imagine life through the eyes of an animal the size of a fly. Conversely, the human world is so large that it is almost imperceivable to the micro-verse. And yet, here we are, co-habiting this planet in our very different ways...

How large unto the tiny fly
Must little things appear! –
A rosebud like a feather bed,
Its prickle like a spear;

A dewdrop like a looking-glass,
A hair like golden wire;
The smallest grain of mustard-seed
As fierce as coals of fire;

A loaf of bread, a lofty hill;
A wasp, a cruel leopard;
And specks of salt as bright to see
As lambkins to a shepherd.

WALTER DE LA MARE (1873–1956)

Vampire Moths

Calyptra spp. | Lepidoptera / Erebidae

Only the males drink blood, the females prefer soft fruit.

No matter how much we learn about the extraordinary diversity of the insect world, it is still capable of delivering the most unlikely surprises. Within a group of insects that are almost exclusively vegetarian, there is a moth that has thrown the rule book out of the window. *Calyptra* – numbering less than 20 species – is the only known genus of moth that drinks blood and, even more strange, it is only the males that are sanguivorous. This is highly unusual, as it is usually females that require blood meals in order to activate the development of their eggs, and it is unknown if males synthesize blood for reproductive purposes. Both sexes have a piercing proboscis; males can pierce the skin of large mammals, whereas females appear to stick to sucking the liquid from soft fruit.

Great Diving Beetle

Dytiscus marginalis **| Coleoptera / Dytiscidae**

The great diving beetle of Europe and Asia can grow to 35mm long.

In temperate regions there are slim pickings for finding adult insects in the middle of winter, but freshwater is still excellent, and one of its most impressive inhabitants is the great diving beetle. Sometimes you can spot it through the ice on ponds. This beetle has long, oar-like hind legs, which are flattened and fringed with hairs to help it move through the water. To breathe, it holds an air bubble on its ventral surface. It is a voracious predator in both the adult and larval stage, and it is perfectly capable of eating fish, frogs and newts. It can also fly from pond to pond, always at night.

FOSSILIZED WASP

Palaeovespa florissantia **| Hymenoptera / Vespidae**

This image is not a painting, but one of the earliest known fossilized impressions of a social wasp, dated to around 40 million years ago.

A picture of a wasp is rendered on a rock. It is exquisitely detailed, such as the impression a screen print or photogram could leave behind. Every abdominal segment is marked out; thoracic marks, even the thinnest veins on the wings are present. But this isn't a picture, it is a fossil. The Florissant Fossil Beds of Colorado, USA, house one of the richest arthropod fossil reserves yet discovered on Earth. The fossil beds are shales formed around 34-44 million years ago, resulting in layers of fossilized organisms, including this beautifully preserved *palaeovespa florissantia*, a direct ancestor of modern hornets and social wasps. This and other wasps were living socially, in colonies, as they still do, and nectaring on flowers that must have been present in the area. Fossils such as this are an extraordinary snapshot of a time otherwise unrecorded, helping us to piece together how the world looked so many years ago.

Orchid Mantis

***Hymenopus coronatus* | Mantodea / Hymenopodidae**

Above: Not only does this mantis look like an orchid, it smells like one too!

Opposite top: The Silverfish is the living fossil in your kitchen.

Opposite bottom: The hump earwig makes a bigger sacrifice than most.

The mantids are famous for their striking prayer posture, and for incidences of post-coital mariticide. Mantids use camouflage to hunt, and many species are slender and green, blending into vegetation. However, the flower mantises have evolved to mimic inflorescences. The orchid mantis is pearly white with subtle markings, and its back two pairs of legs are modified to look like the petals of orchids. Although it appears to just mimic the orchid's appearance, its success could be down to more complex factors. Smelling like, and copying the UV markers of, an orchid may also be at play in the mantis' strategy for luring its prey – so much so that controlled experiments have found that the mantis is actually more attractive than its floral counterparts!

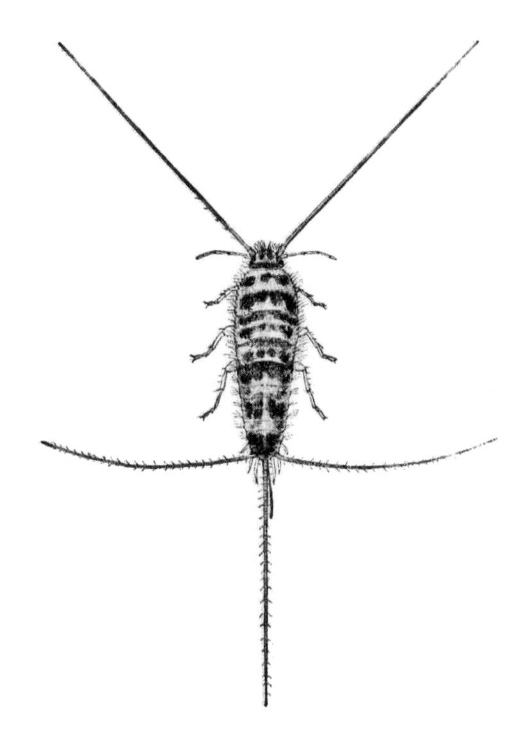

6TH JANUARY

Silverfish

Lepisma saccharinum

Zygentoma / Lepismatidae

If you've ever glimpsed a silverfish scuttling away when you switch on the light in your kitchen, you have looked through a window into the distant past. The silverfish is a small, flightless insect that closely resembles those that dominated in the Silurian period 420 million years ago, before they developed wings. It is thus a throwback, and a survivor.

7TH JANUARY

Hump Earwig

Anechura harmandi

Dermaptera / Forficulidae

Mothers the world over are famous for the sacrifices they make for their offspring, and they number insects among them. Few make the sacrifice quite as fully as the hump earwig, however. This insect breeds in the winter, which is great for avoiding predation, but not so good for gathering provisions to feed the growing nymphs. The answer is known as matriphagy – the offspring kill and eat the mother.

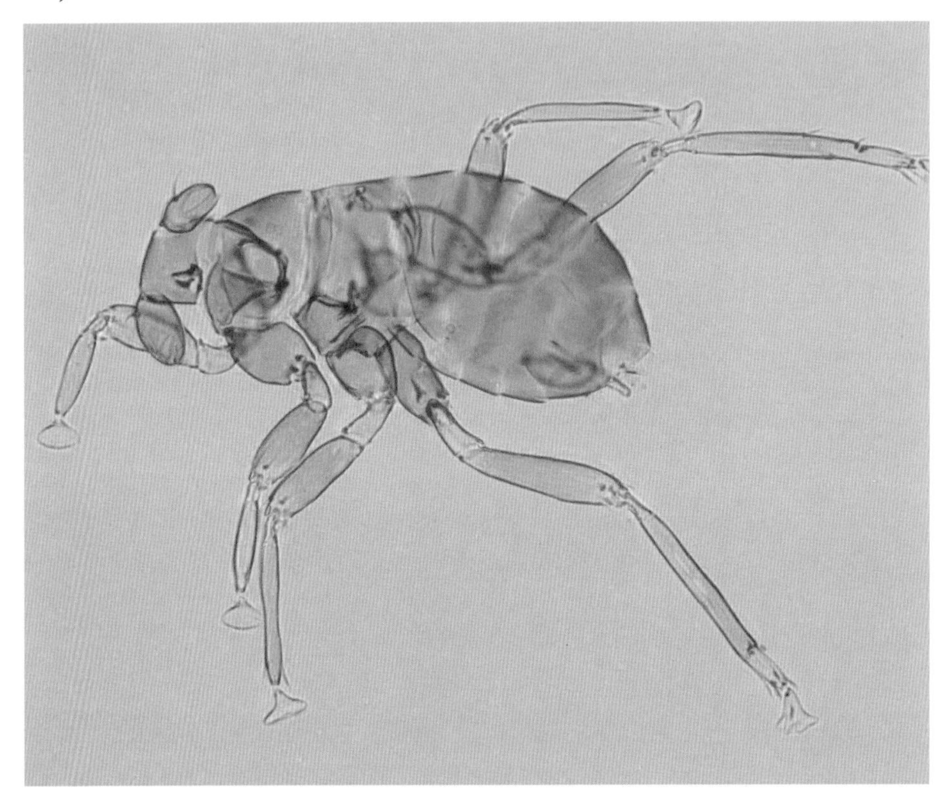

Fairy Wasp

Dicopomorpha echmepterygis **| Hymenoptera / Mymaridae**

The fairy wasp is the smallest insect you can encounter.

Size matters. Or does it? In a world where bigger is better, let's ponder the fact that the extraordinary evolutionary success of insects is largely (no pun intended) due to their size. Being small is a great survival model, but just how far can you physically take that? The smallest insect currently known, *dicopomorpha echmepterygis*, is a wasp that occupies less space than an amoeba. It is about as small as we think functional multicellular life can possibly get. The blind, wingless males measure around 140μm, less than the width of a human hair. The comparatively gargantuan females, at around 200μm, are still so small that their wings do not require a membranous surface. At this scale, air moves more like liquid and the wasp can be propelled along, aided merely by thin paddles fringed with long, delicate hairs. If you haven't heard of these micro-beasts, it isn't because they are rare. On the contrary, they are everywhere; but their near-impossible size means that we just don't see them.

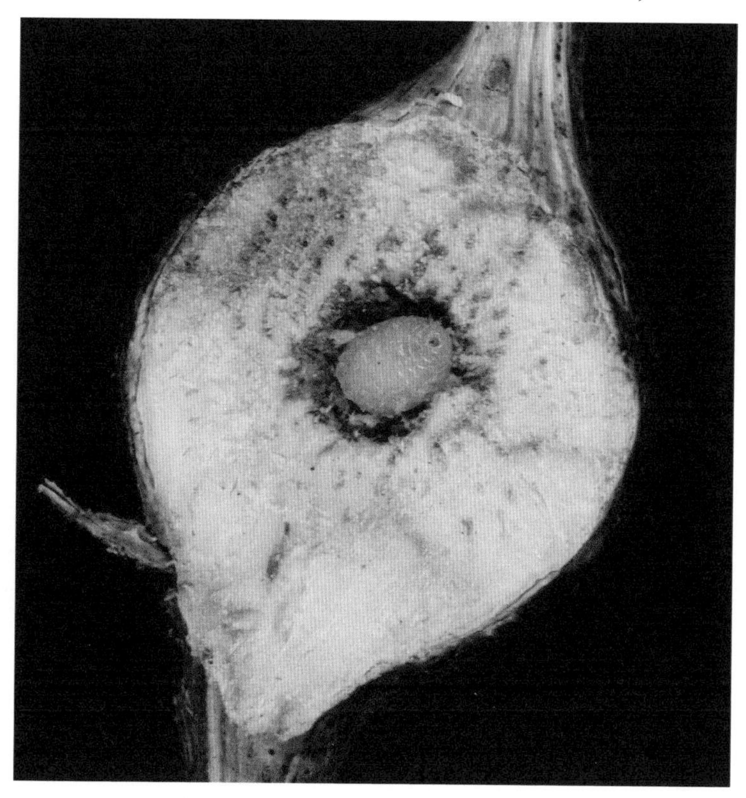

Goldenrod Gall-fly

***Eurosta solidaginis* | Diptera / Tephritidae**

Goldenrod gall-fly larvae can survive partial freezing in the depths of winter.

It's remarkable how many small and seemingly unexciting insects have extraordinary talents. In the case of the goldenrod gall-fly, it's the larvae – they are incredibly hardy, being among the very few insects that are cold tolerant, allowing ice to form in their tissues. The larvae live in galls on the goldenrod plant (*Solidago* spp.). Their tissues contain glycerol and other cryoprotectants, which means that they don't freeze until it is at least –20°C (–4°F). But they can cope beyond that, their body allowing ice to form in the gaps between cells. These animals have ice in their veins – and mild winter temperatures can be fatal to them.

Above: Army ants must carry their larvae across the forest floor when relocating.

Opposite top: Timber flies are gentle giants, and superbly coloured, with bright yellow wings.

Opposite bottom: These butterflies have complex wing structures that help them camouflage in the dappled rainforest sunlight.

Army Ant

Eciton burchellii **| Hymenoptera / Formicidae**

Two hundred thousand ants on the move makes for an intimidating sight. The ground seethes with them as they march in living tributaries across the forest floor, searching for any living thing of the right size to feed their growing larvae back at the colony. During a 12-hour raid, the columns, some 200m (650ft) long, may kill as many as 100,000 other animals as they sweep across an area, leaving the ground bereft of life. The next day, they will do it again. Each night, the ants make a bivouac of nothing more than living bodies, held together with hooks on the workers' feet, in a different place each night.

The daily raids happen during the ant colony's 'nomadic phase', when the eggs have hatched and the larvae are always hungry. After 15 days, the colony enters a 'statary phase', when the queen is in egg-laying mode, and the raids are only once every other day. During this phase, they ants may bivouac in the same place on consecutive nights.

Timber Flies

***Pantophthalmus* spp.**

Diptera / Pantophthalmidae

Timber flies of Central and South America are enigmatic, gentle giants that almost rival the Mydas flies in length and wingspan. Females lay eggs in trees or deadwood. The larvae develop in galleries within the wood; their diet is something of a mystery – possibly fermenting sap or the wood itself. They aren't quiet about it either, and can be heard munching away inside the tree from metres away.

12TH JANUARY

Clearwing

Greta oto

Lepidoptera / Nymphalidae

Common clearwing butterfly wings have two distinctive scale structures. The clear area has waxy, anti-reflective properties and sparse, hair-like scales that transmit light. The coloured parts have densely packed, tooth-shaped scales that not only block and reflect light, but also prevent the wings sticking together. The result is a living trompe-l'oeil in the forests of Central America.

Arctic Tern Louse

Quadraceps houri **| Psocodea / Philopteridae**

The much-travelled Arctic tern carries passengers with it.

What's the world's best-travelled insect? The answer is almost certainly an insect that cannot actually fly. If you discount headlice, which could be attached to well-travelled humans (even astronauts!), then the lice or fleas of birds would be good candidates.

One strong possibility is a parasite attached to a long-distance migrant, such as the Arctic tern (*sterna paradisaea*). *Quadraceps houri* is an example, which uses hosts of several bird species. The Arctic tern flies from the Arctic to the Antarctic in autumn and back again in spring, and may cover 80,000km (50,000 mi) a year. The average bird louse lives about one month, so that could still be quite a ride!

Leafcutter Ant

Atta spp. | Hymenoptera / Formicidae

Humans don't have a monopoly on farming; leafcutter ants cultivate fungal hyphae as food for themselves.

By the time humans even thought about cultivation and farming, the leafcutter ants of South America already had a long-established, highly successful system of agriculture that still exists today. Over 200 species of leafcutter ants have developed a mutualistic relationship with fungi, which they grow and harvest within their nest for food. You've probably seen documentary footage of lines of ants carrying sections of leaf in convoy across the forest floor. Perhaps surprisingly, the ants don't eat the leaves themselves; back in the nest, they chop up the leaves and spread them over the subterranean fungal threads called hyphae. The hyphae ingest the leaves and are themselves eaten by the ants. But why go to all that effort, when the ants could simply eat the leaves? The plant material contains polymers that the ants cannot metabolize, meaning they cannot access all of the nutrients within. However, the fungi can break down those polymers, and by eating the fungi the ants gain significantly more nutrients. This mutualistic relationship has been happening for so long that the cultivated fungi found in ant nests cannot be found anywhere else in the wild; so specialized is it that a newly emerged queen will take a 'cutting' of her nest's hyphae with her when she leaves to begin her own colony.

Western Honeybee

Apis mellifera mellifera **| Hymenoptera / Apidae**

It may come as a surprise that the honeybee has the ability to count.

We celebrate the honeybee today, clever insect extraordinaire.

In recent years, scientists have unearthed abilities and attributes that, until recently, would have seemed inconceivable for an insect with a tiny brain of only 1 million neurons (by comparison we have 100 billion). Who, for example, would ever have predicted that an insect could count? But they can, to four (it takes humans at least 10 months to acquire this ability). Recent research even suggests that they can do simple sums. They can also remember the colours and shapes of flowers for several days (but they cannot see red colours). They can work out complex tasks such as lifting lids to get food. They can also pass on instructions to other bees and, in turn, learn how to do things by watching demonstrations.

Honeybees are not little robots, but have distinct personalities, with individual strengths and faults. Some are not as effective at the tasks they are meant to perform as others. Some, you may be delighted to hear, just aren't that good at housework!

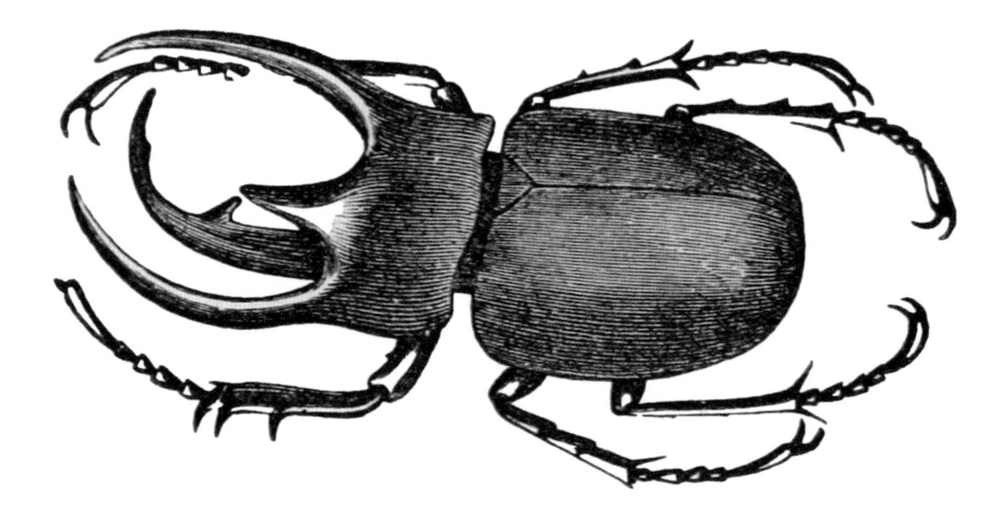

An Abundance of Beetles
Coleoptera

Above: The beetle body blueprint is one of the most successful on Earth.

Opposite top: The only way to get these flies facing sideways is to pin them!

Opposite bottom: The ice crawlers have adapted to the harshest conditions on Earth, surviving at temperatures that are almost permanently sub-zero.

With at least 350,000 described species, the order Coleoptera outranks every other known group of animals on the planet. One in four of every described species of anything is a beetle, and a group of beetles, the weevils (Curculionoidea), contain more living species (80,000) than all birds, mammals, reptiles, amphibians and fish put together.

'The Creator, if he exists, must have an inordinate fondness for beetles.'

J. B. Haldane

'Whenever I hear of the capture of rare beetles, I feel like an old war-horse at the sound of a trumpet.'

Charles Darwin

'It seems therefore that a taste for collecting beetles is some indication of future success in life.'

Charles Darwin

17TH JANUARY

UPSIDE-DOWN FLY

Neurochaeta inversa

Diptera / Neurochaetidae

With a name like that, these flies simply have to occur in Australia, don't they? Only discovered in the 1960s, they are now also known to be from New Guinea, Malaysia and Africa. They have a very strange habit of always clinging on to a vertical surface with their head pointing down. If you put them in a container and turn it upside down, all the flies reorientate to face down. Whatever direction they go, even sideways, the head stays down. If they ascend, they walk backwards.

18TH JANUARY

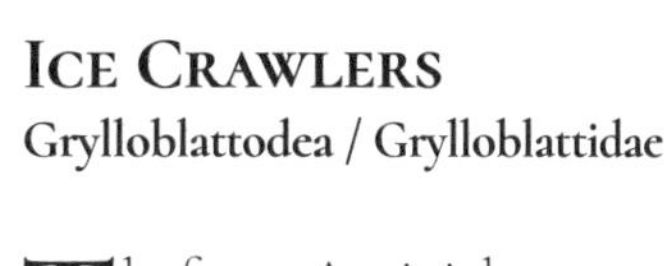

ICE CRAWLERS

Grylloblattodea / Grylloblattidae

The frozen Arctic is home to some highly specialized insect species, including the ice crawlers. A rather unique group of insects, they look like a mixture of termite, cricket and cockroach. Ice crawlers endure almost constantly freezing temperatures in the Arctic Circle, living in ice caves or underground, feeding on any dead organic matter they can find. Their unusually slow metabolism means that they can also live much longer than many insects – up to ten years.

Blushing Phantom

Cithaerias pireta | **Lepidoptera / Nymphalidae**

The blushing phantom lives a twilight existence in the shadowy rainforest understorey.

Butterflies are well known for their habit of flying in the daytime and loving warm, sunny weather, in contrast to their relatives, the moths, which are generally nocturnal. Nature does, however, like to bend the rules. We have moths that fly by day, and there are also butterflies that like the dark. The Haeterini are a tribe of butterflies, including the blushing phantom, that live in the dense understory of the Amazon rainforest. They negotiate their way through the gloom in a leisurely fashion, feeding on rotting fruit and fungi. These elusive butterflies are not easy to find in the almost permanent twilight between forest floor and canopy, and they emerge at dusk, possibly to avoid diurnal predators. Their inconspicuous behaviour is compounded further by their morphology: Haeterini wings are a delicate mix of pigment, structural colour and translucency, the result of which is a diaphanous film which reduces them to mere phantoms in the shadows.

Saltpool Mosquito

Opifex fuscus **| Diptera / Culicidae**

The saltpool mosquito lives in rock pools on New Zealand's coast.

This mosquito is endemic to New Zealand, where it lives by the ocean inhabiting coastal rockpools and intertidal margins (only about 5% of mosquito species live in this type of habitat). Eggs are laid above the waterline and are washed into rockpools where they overwinter; the emerging larvae are omnivorous, eating small invertebrates and tidal sediments. Saltpool mosquitoes are highly unusual in that adult females do not need to ingest blood for their eggs to start developing.

GLOWING CLICK BEETLE

***Deilelater physoderus* | Coleoptera / Elateridae**

The glowing click beetle can produce a bioluminescence so bright that it looks like it is equipped with headlights.

As if click beetles aren't fabulous enough, with their nifty, spring-loaded hingelock system that propels them through the air with a G-force that can exceed 300, some of them have their own headlights, too! The glowing click beetles of the American tropics are bioluminescent. The back corners of the pronotum shine brightly with the presence of oxygen, adenosine triphosphate (ATP) and the bioluminescent enzyme luciferin, which, when combined, produce a light that is bright enough to be seen at some considerable distance, which has has earned them the nickname 'headlight elaters'. They can even control the intensity of the light with an in-built 'dimmer switch'.

Milichid Flies

Phyllomyza spp. | Diptera / Milichiidae

The Milichid fly's small size gives it unusual feeding opportunities.

'Will you walk into my parlour?' said a spider to a fly;
''Tis the prettiest little parlour that ever you did spy.
The way into my parlour is up a winding stair,
And I have many pretty things to shew when you are there.'

'Don't mind if I do,' said Phyllomyza, 'kind of you to ask.
'It happens that I'm sitting here and ready for the task.
'You cannot see me, I'm too small, my body you can't feel,
'But when you've wrapped the struggling prey, my purpose I'll reveal.'

'When work is done, when enzymes sweet have bodily digested,
'The fly right here, whose sad demise your silken web invested,
'I'll play my hand and drink some juice, I only need the dregs,
'And when I'm full I'll leave, unseen, to go and lay my eggs.'

Adapted from the poem 'The Spider and the Fly' by Mary Howitt (1799–1888)

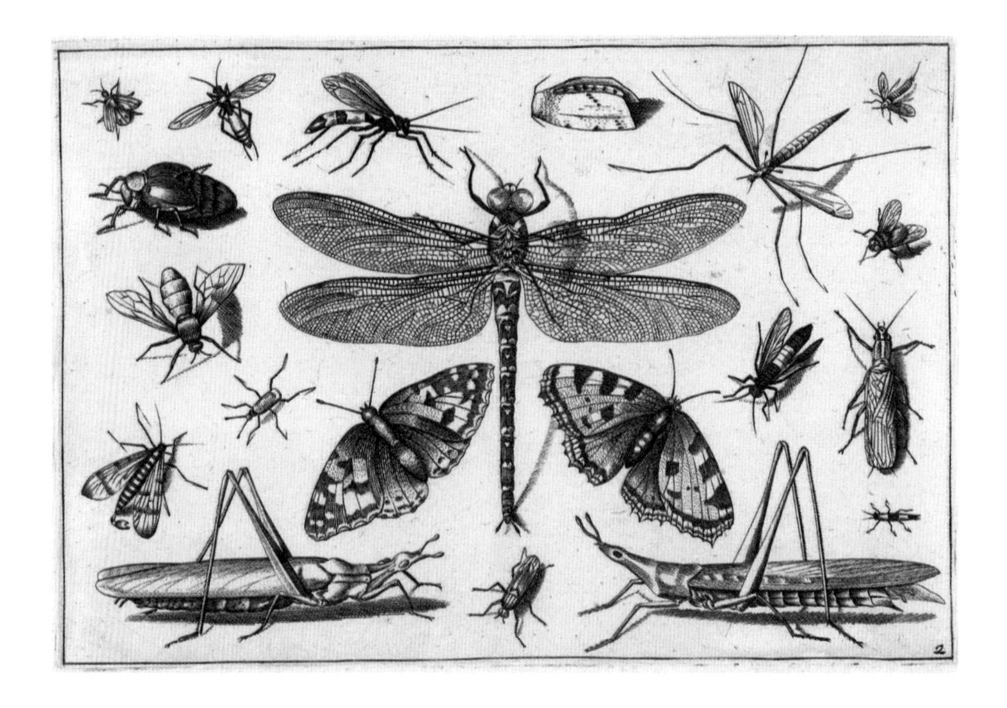

Wings

Insect wings have evolved over hundreds of millions of years into vastly diverse structures that suit many different lifestyles and purposes.

Apart from birds and bats (and prehistorically the pterosaurs), insects are the only group of animals on Earth capable of flight. No other animals have successfully evolved parts of their bodies into appendages that can carry their body weight through the air for long periods. The possession of wings gives insects an enormous advantage in the game of life, as flight allows you to travel greater distances to find food and potential partners. A further contribution to success are the different ways in which insect wings have adapted over millions of years, creating distinct groups. In flies, for example, the hindwings have been reduced to small appendages that act as flight stabilizers, leaving the forewings for actual flight. The beetles have adapted their forewings into hardened casings – often colourful or patterned – that conceal the flying wings. And wings aren't only used for flight; they can also be a powerful communicator during courtship displays, or to establish territory.

Glitter Weevils

***Pachyrhynchus* spp. | Coleoptera / Curculionidae**

These spectacular, glittery weevils have to be seen to be believed; they're part beetle, part kid's art project!

Sometimes an insect comes along that just doesn't seem real. The almost ubiquitous usage of image manipulation and AI can have us doubting whether what we have seen can possibly be genuine. This can particularly be the case for many insects, some of which appear in this book and prove that fact is truly stranger than fiction. And then there is the glitter weevil. It doesn't look like the pinnacle of photoshop though; it looks like a child stuck sequins all over the family dog. This weevil has a rich, black cuticle that is adorned with the most spectacular, tiny iridescent scales. It is the sort of experimental crafting that may initially bring out the cynic in you, and have you wondering what random turns evolution felt the need to take to end up with a miniature disco pug. But once the disbelief has diminished, you will realize that this is not a TikTok prank, but a genuine animal better than anything the human imagination could concoct.

Metallic Stag Beetle

***Cyclommatus metallifer* | Coleoptera / Lucanidae**

This is the gold form of the extraordinary metallic stag beetle.

There are impressive stag beetles in Europe and the USA, but for sheer star quality, both are outshone by this extraordinary beetle from Indonesia. It comes in two colour forms, gold (which is the most common) and black, which naturally arise in both wild and captive populations. The males bear the outsized mandibles, which are used in combat over females. Well-matched males indulge in gladiatorial combat, jostling with their mandibles and trying to bite one another, with an end game of throwing your rival off the log or branch on which you are competing.

Leaf-roller Fly

Trigonospila brevifacies | **Diptera / Tachinidae**

The leafroller fly is a super-smart fly native to Australia and New Zealand.

Parasitic flies may not sound particularly endearing to most people, but here's one that surely cannot fail to charm. The leaf-roller flies (*trigonospila* spp.), found in Australia and New Zealand, are snazzily striped species with crisp black and white (or pearly) horizontal bands, making them look like tiny, walking pedestrian crossings. These enigmatic flies are thought to parasitize leaf beetles (chrysomelidae) and for this reason are seen as beneficial biological control.

Spur-legged Stick Insect

Didymuria violescens

Phasmatodea / Phasmatidae

This is a common species of stick insect, which, bearing in mind it lives in Australia, has made the very sensible career move to feed on the leaves of eucalyptus. As these are the dominant local trees, numbers of this phasmid can easily build up to almost plague-like proportions, and they can defoliate entire forests. Strangely, the more abundant they get, the more colourful they become, seemingly abandoning their usual green and brown camouflaging colours. The flying males, in particular, can be gorgeous with their startling violet wings.

Mealworm Beetle

Tenebrio molitor **| Coleoptera / Tenebrionidae**

Above: Beetles and other insects are already intrinsic to the diet of many countries and could be the key to future global food security.

Left: This stick insect shows off impressive colours in flight.

Did you know the mealworms you leave out for your garden birds are actually a beetle? Mealworm beetle larvae are bred in farms, then dehydrated and packaged into a wide variety of animal feed, including bird feed. They have a high percentage of protein relative to their mass, making them extremely energy efficient to both produce and to eat. They require less water, space and food, and produce less CO_2 per kg of bodyweight than cattle or similar livestock. The European Food Safety Authority recently classified mealworms as safe for human consumption, and around the world they are already an integral part of our diet. Ground-up mealworms are made into a protein powder that can be incorporated into everything from milkshakes to burgers. These beetle larvae are likely to become a standard Western ingredient on a par with large mammal livestock and could be the key to our planet's food security.

DANCING JEWEL

***Platycypha caligata* | Odonata / Chlorocyphidae**

It's the white flashes on the legs of the male dancing jewel that catch a female's attention.

Damselflies, those high-viz insects that live in a world of dazzling colours, go in for some pretty fervent, highly charged courtship. In the case of the dancing jewel, there is more leg-showing than a tango. The male is adorned with a fancy livery, with a shining, powder-blue abdomen, plus crimson below the abdomen, and long legs with an eye-catching combination of red and brilliant white. As soon as a female approaches, the male flashes his white tibiae, then flies around the female and waves his blue abdomen, without subtlety. If the potential mate remains in place, the zenith of the display is to hover in front of her, shivering the white parts of the legs until they are a blur.

Honey Ant

***Camponotus inflatus* | Hymenoptera / Formicidae**

Honey ants have evolved a novel way to ensure year-round food availability, by turning part of their workforce into a living larder.

In the Australian desert, where water and food are scarce, a group of ants has taken matters into their own hands, literally turning themselves into a food store for the rest of the colony. Honey ant colonies contain a section of workers that have incredibly elastic abdomens, within which a rich, sugar-rich liquid is stored. These workers hang from the ceiling of the nest while provisions of honeydew are delivered and fed to them by foraging sisters. When fed, the abdomens of these 'living larders' distend to huge proportions, bursting with sugary liquid. They remain in place like little vending machines, regurgitating their honey stores, as required, to feed the colony in lean times. Foraged by Indigenous people for generations, their 'ant honey' is recognized as having antimicrobial properties, the applications of which are already used by communities to heal wounds and soothe throat infections – yet another way in which insects are beneficial to humans.

Social Wasp

Polybia paulista **| Hymenoptera / Vespidae**

Far from being our enemy, this social wasp (*Polybia paulista*) could in fact save many of us in the fight against certain cancers.

It's safe to say that humans aren't keen on being stung by wasps. The venom delivered into the skin, depending on the wasp, can register anywhere from mild annoyance to blinding pain. *Polybia paulista* is a social wasp that is endemic to south-east Brazil. It is one such wasp capable of administering a painful injection when feeling particularly threatened itself, or on behalf of its nest colony. But the venom of this wasp contains something else that we have discovered is far more beneficial. *Polybia paulista* contains a peptide that has been named Polybia-MP1 (MP1 for short) and is capable of selectively killing some cancer cells. This wasp is just one example of how our problematic relationship with insects may actually be causing us more harm in the long term. We can potentially unlock the key to many more medical breakthroughs, if we focus our efforts on conserving and understanding insects that pose a risk to us, rather than seeking to eliminate them.

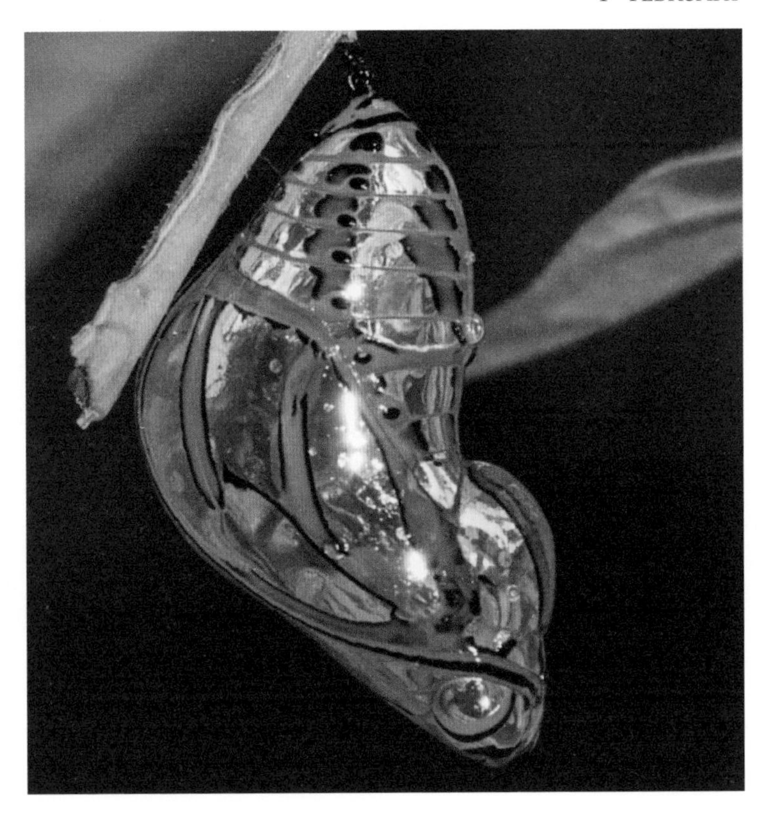

Harmonia Tiger-wing

***Tithorea harmonia* | Lepidoptera / Nymphalidae**

The Harmonia tiger-wing chrysalis, with its mirrored gold surface, looks more like a Christmas tree ornament than a pupa.

The Harmonia tiger-wing begins life as a fairly modest caterpillar which must – like every butterfly – embark on the complex process of pupation. The majority of butterflies will temporarily transform into a lumpen mass of green or brown for a few weeks, to blend into vegetation or soil respectively. This butterfly, however, plays a very different game; its pupa is gold. A combination of pigment, reflective structure layers and layers of fluid create a mirror-like surface that you can literally see yourself in. The highly reflective nature of the pupae could be cryptic camouflage, masking it as a drop of water, or breaking up its outline in the visible spectrum of predators. The shell of the pupa is transparent, and close to eclosion the golden colour disappears and the brown, red and white colours of the fully formed adult butterfly are visible through the clear outer layer.

Brown China-mark

***Elophila nymphaeata* | Lepidoptera / Crambidae**

Above: The Brown China-Mark is a moth with a very unusual life cycle.

Opposite top: Cold temperatures don't bother the Winter Ant.

Opposite bottom: The Yellow-faced Bees of Hawaii have evolved into multiple species.

The insect world is full of anomalies; some species in a taxa adapt in new ways and follow very different paths to their relatives. Take the brown China-mark, for example. It looks like a moth and behaves like a moth. As an adult, that is. It is a little unusual in that it flies by day, but a surprising number of moth species do this. But the really interesting thing about this moth is the larva – it is aquatic; a caterpillar that lives in water. This is highly unusual for moth larvae, as we know them to be tree or herb-based eating machines. This caterpillar however, feeds on water plants. It lives within leaf tissue in its early days, then cuts out a section of leaf, folds it around itself like a pitta pocket, and floats about in it as it matures.

WINTER ANT
Prenolepis imparis
Hymenoptera / Formicidae

This remarkable ant avoids competition by being out and about when other ants are tucked away underground, in the very depths of winter. Only when the weather is cold and grey does it come out foraging, often when it's freezing. These ants have been observed moving around when it is –2.7°C (27.1°F), and when there is frost on the ground. In the summer they shun the surface, living as much as 3.6m (almost 12ft) down, where they undergo a kind of aestivation.

4TH FEBRUARY

NALO MELI MAOLI
Hylaeus **spp.**
Hymenoptera / Apidae

Hawaii is one of the most remote places on Earth, 3,200km (almost 2,000 miles) from North America and 9,000km (5,600 mi) from Australia. 500,000 years ago, a yellow-faced bee made landfall here from one of these continents and flourished in the absence of any other bees. It evolved into 60+ endemic species. Several are now disappearing, threatened by fire, habitat loss and invasive ant predators.

Green Tree Ant

Oecophylla smaragdina **| Hymenoptera / Formicidae**

Green tree ants work as a team to nest build, using their own larvae as glue sticks to bond leaves together into a cosy canopy condo.

In parts of Australia and Southeast Asia, you may find a peculiar adornment hanging from a branch. Green tree ants (also known as Asian weaver ants) are arboreal; rather than nesting in the ground, they construct large, intricate baubles from leaves, with the help of their larvae. A group of workers curl the leaf around, pulling the sides together to meet each other. Another unit of workers is already on standby, and each of them holds a larva. As the leaves are pulled together the larvae-brandishing workers begin. The larvae spin sticky silk from the tips of their abdomens, and the workers use this silk to bond the leaf edges together. Hereupon, the larvae are effectively turned into glue sticks, their little rear-ends pressed together to stimulate silk production. The adhesion process is highly efficient; the end result is a beautiful leafy sphere that will house eggs and larvae. A single colony will produce multiple nest spheres across multiple trees, creating a kindergarten metropolis in the canopy.

Mammoth Wasp
***Megascolia maculata* | Hymeanoptera / Scoliidae**

Unsurprisingly the mammoth wasp is the largest in Europe.

For wasp lovers, few experiences can match the sheer excitement of spotting the Mammoth Wasp as it busily goes about its daily duties. It will be unaware of the kerfuffle it is causing among fascinated onlookers, for this is Europe's largest wasp. It belongs to the Scoliidae; a family of parasitoidal solitary wasps comprising 560 species globally, the heftiest of which are the *Megascolia* genus, to which this beautiful, buttercup-headed one belongs. Scoliid wasps are big and robust because they hunt the substantial larvae of scarab beetles; a female digs into a beetle burrow, grabs and stings a larva, paralysing it with venom. It then injects the larva with one of its own eggs, which will hatch and consume the larva from the inside. Adult scoliids can be seen nectaring in the sun, minding their own business; though fearsome in reputation, these wasps will only sting larger animals as last-resort defensive behaviour.

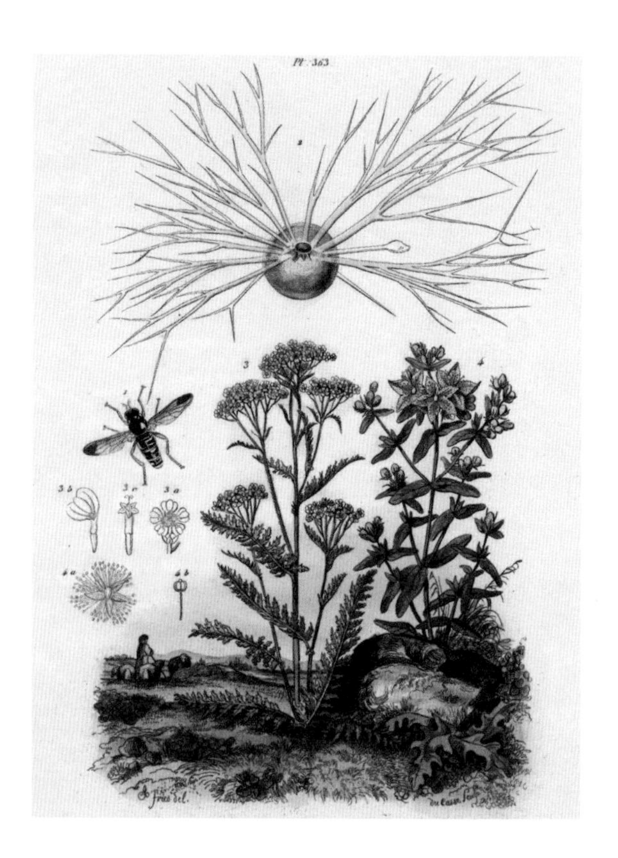

Yellowjacket Mimic

Spilomyia longicornis | Diptera / Syrphidae

Strong yellow and black markings fool potential predators into thinking that this harmless fly is a dangerous wasp.

You would be forgiven for thinking that this *spilomyia longicornis* is a wasp. It is actually a fly – a hoverfly to be more specific. The telltale signs are the thick waist (much thinner in wasps) and the short, stubby, three-segmented antennae. It also has just one pair of flight wings (wasps have two); the hindwings have evolved into much smaller, gyroscopic appendages, called halteres, that aid balance and movement through the air. But other than that, the colour, pattern and posture are all distinctively waspish. It is a textbook example of Batesian mimicry, in which a harmless prey species evolves the appearance and mannerisms of a more dangerous predatory species, in order to improve its own survival prospects.

Isabella's Longwing

Eueides isabella | Lepidoptera / Nymphalidae

Heliconid butterflies are a familiar sight in the American tropics.

The Isabella's longwing is in a group of butterflies often called heliconids, which have extended forewings and a distinctive shape and flight. They are abundant in the tropics, especially in South America. They have several unusual habits. One is that their caterpillars eat extremely poisonous flowers, in this case passionflowers (*Passiflora*). Secondly, males often attach themselves to the pupa of a female before she has time to hatch. And thirdly, they often roost in groups at night, flying until it is almost dark.

Sacred Scarab

***Scarabaeus sacer* | Coleoptera / Scarabaeidae**

This amulet would have been carried by an ancient Egyptian with the belief that it would protect them or bring them good fortune.

In polytheistic ancient Egypt, Khepri was the sun god of dawn, an aspect of the greater sun god, Ra. The name Khepri is derived from the meanings 'to create' or 'to come into being' and symbolizes the rising sun and coming of morning; Khepri was believed to steer the sun (or Sirius, depending on the account) across the sky on its journey from dawn to dusk. Parallels were drawn between this belief and the behaviour of the scarab beetle that inhabited the coastal dunes of Egypt and the southern Mediterranean. Scarabs are dung collectors, and they do this in the most charismatic way. Using their crimped heads and front legs, they hack away at the dung of herbivorous ungulates, then they roll the bounteous matter into balls and make away with it to their nursery. These perfect spheres can be considerably larger than the beetles, and so the most efficient translocation method involves the beetles pushing the dung balls backwards with their hindlegs, which they perform with admirable speed and efficiency. For ancient Egyptians, the observation of the scarabs, rolling large balls across the ground became analogous with Khepri's celestial shunting of the sun across the ecliptic, and so his earthly form became a human with the head of a scarab beetle. The scarab itself came to adorn jewellery, clothing, seals and amulets, becoming one of the most familiar motifs of ancient Egyptian symbolism.

Giant Wood Moth

***Endoxyla cinereus* | Lepidoptera / Cossidae**

Giant wood moths spend the majority of their lives concealed within the living wood of trees, and can create quite a stir when they emerge as particularly large adults.

Native to Eastern Australia and New Zealand, the giant wood moth is the heaviest moth in the world. The largest females can weigh in at an impressive 30g (1oz), reaching 15cm (6in) in length and with a wingspan of up to 25cm (10in); males are generally about half the size. They have an enigmatic life cycle; adult moths are greyish-brown in colour with darker spots on the wings, which helps them blend into the bark of their host tree, the eucalyptus (within the crevices of which up to 20,000 eggs are laid). The newly hatched larvae, commonly known as 'witchetty grubs' are thought to lower themselves to the ground on silken threads and burrow into the soil to feed on roots for the first year. They then crawl back up the trunk and bore a hole into the wood, burrow in and spend at least another year or two maturing in preparation for pupation. The adult phase is fleeting – just days or weeks, and in this time they will not feed, focusing only on finding a mate. Despite their size, adult giant wood moths are infrequently seen, but when there are, they cause quite a stir.

FIREBUG

Pyrrhocoris apterus | Hemiptera / Pyrrhocoridae

It seems that personality could be important to firebugs.

Insects have a constant capacity to surprise. Who would have thought that a bug whose main aims in life are to suck sap and mate would have the subtlety to show off different individual personalities? Yet the firebug does – and so do many other insects, even with their minute brains.

Firebugs have been subject to some intriguing experiments. A number of individuals released into an arena with unfamiliar objects (such as bottle caps) show strikingly different degrees of inquisitiveness. Some stay well away from the unfamiliar, while other individuals embrace it. Some are simply more active than others.

In another experiment, their sociability was tested and, sure enough, there were stark individual differences. Of course, all this makes sense. If everybody is the same, everyone could make the same mistake. Variety is a good survival strategy.

The Insect Brain

The great English scientist Charles Darwin was born on this day in 1809.

'It is certain that there may be extraordinary mental activity with an extremely small absolute mass of nervous matter: thus the wonderfully diversified instincts, mental powers, and affections of ants are notorious, yet their cerebral ganglia are not so large as the quarter of a small pin's head. Under this point of view the brain of an ant is one of the most marvellous atoms of matter in the world, perhaps more so than the brain of a man.'

Charles Darwin, (1809–1882)

Tarantula Hawk Katydid

***Aganacris* spp. | Orthoptera / Tettigoniidae**

Despite its scary, wasp-like appearance, this animal is a harmless bush-cricket with a beautifully evolved \.

In the dense greenery of the tropical rainforests of South America, you suddenly find yourself face-to-face with a large, velvet-black insect with orange wings and a bulbous abdomen. Every alarm bell in your body and brain goes off simultaneously; a primeval adrenalin spike triggers your flight response, because you are in immediate danger. The tarantula hawk in front of you could dispense a sting of unbelievable volatility. You slowly back away, praying it doesn't hear your pounding heart or smell the fear oozing from every pore. The relief at surviving such a close encounter has you laughing – or crying – with relief. But the joke is on you because this isn't a tarantula hawk at all; it's a harmless bush-cricket.

LOVEBUG

Plecia nearctica **| Diptera / Bibionidae**

A romantic misfire is fatal for the lovebug.

Here is a tale of fatal attraction. Lovebugs are small black and orange flies that emerge en masse in a nebulous cloud, but things take a bizarre turn, because lovebugs are hopelessly attracted to... roads. Studies have found that lovebugs are irresistibly drawn to a combination of exhaust fumes, UV levels and warm temperatures (28°C or above). They congregate amid this heady cocktail, oblivious to the danger they are in. The result of this romantic misfire is a very different coming together – that of vehicle and lovebug, and this can happen on a huge scale. Automobiles are caked in the carcasses of these small flies and yet, in a plot twist worthy of any romantic comedy, the lovebug has the last laugh, as it sets like dried porridge on paintwork, and if not cleaned off immediately, eats into the paint, leaving behind a lasting mark; it is the stuff of drivers' nightmares and carwash owners' dreams.

Snake-Caterpillar Moth

Hemeroplanes triptolemus **| Lepidoptera / Sphingidae**

Look out, it might bite!

Many insects throughout the world have evolved marvellous ways of keeping predators at bay. Hats off, though, to the snake-caterpillar moth, which really has taken its tribute act into Oscars realm. While the adult is quite unremarkable, its caterpillar has evolved its rear end to look exactly like the front end of a snake – no doubt the vine snake that shares its habitat.

This caterpillar doesn't just sit there and let you admire it; there is some play-acting to come. It extends its rear end and even shakes it, to pretend that it is striking at you.

TANGLE-VEINED FLY

***Moegistorhynchus longirostris* | Diptera / Nemistrenidae**

This species of tangle-vein fly are endemic to the west coast of South Africa.

Although bees are commonly acknowledged as being our most important pollinating insects, the humble flies actually deserve much more credit for making our world go round. One of the earliest pollinator groups on Earth, they have evolved to fill almost every nectar-guzzling, pollen-shifting niche out there. Some flowers are so unusually shaped that drastic evolutionary measures have been taken by many flies, including *Moegistorhynchus longirostris*, a species of tangled-vein fly from the west coast of South Africa, which has a ridiculously long feeding tool. An important pollinator of a number of endemic plant species, this fly feeds just like a hummingbird, by inserting its extensive proboscis (which can reach around 9cm/3½in in length, the longest relative to body size of any known insect) into the most inaccessible flower heads of orchids, irises and pelargoniums.

Long-Snouted Weevil

***Brentus anchorago* | Coleoptera / Brentidae**

Above: Males of this species vary greatly in body length.

Opposite top: The Patagonian bumblebee is so large it looks more like a flying mouse.

Opposite bottom: Male and female leafhoppers dance to a different beat.

We all know that we live in an unequal world, but the world of this South American weevil is more unequal than most. This species must surely hold the record, at least among insects, for the largest recorded difference between the sizes of individual adult males. The shortest males measure 7mm (¹/₃in) in length, while the longest can reach as much as 52mm (2in), seven times as long. In terms of weight the discrepancy is greater still, with the largest males being up to 26.5 times heavier.

18TH FEBRUARY

Patagonian Bumblebee

Bombus dahlbomii

Hymenoptera / Apidae

The Patagonian bumblebee is also known as the flying mouse, on account of its size (at nearly 3cm, it is one of the largest bumblebees on Earth) and its dense fur, which lends it a mammalian appearance. Once a common sight throughout South America, it is now threatened with extinction, following the introduction of smaller, non-native bumblebees imported for the purposes of intensive pollination of agricultural crops.

19TH FEBRUARY

Two-marked Treehopper

Enchenopa binotata

Hemiptera / Membracidae

Male and female two-marked treehoppers are well-dispersed, so initial contact is by a series of 'text messages'. A male taps a stem hard and quickly, sending a vibration through the plant. If a female picks this message up through her legs, she responds by tapping, but at a different frequency.

Snow Fleas

***Boreus* spp. | Mecoptera / Boreidae**

Snow fleas don't just come out in the snow, but you are more likely to see them as their dark forms scuttle across the sparkly white carpets.

During the colder months, when the ground is frozen, a tiny and remarkable creature can be found scampering around in blankets of moss in the boreal regions of the Northern Hemisphere. Snow fleas are members of the Mecoptera – the ancient order of insects to which scorpionflies also belong; the similarity being clear in that long, rostrum-shaped head. However, unlike their vernal relatives, snow fleas emerge in temperatures that would test the hardiest of humans. They are so well adapted to the cold that they are unable to cope with warmer conditions and will struggle when temperatures reach around 16°C. They are small (around 5mm), dark and unassuming, so the best time to spot them is after fresh snowfall, when they show up against the bright white background. They cannot fly; females are wingless, and the males have reduced wings, so their primary method of locomotion is to scuttle about as they search for food such as moss or small invertebrates.

Termite Undertakers

Reticulitermes fukienensis **| Blattodea / Rhinotermitidae**

Termites are famous for their caste system, in which individuals carry out certain assigned tasks.

Insects living in large colonies are particularly prone to the transmission of disease, and in the case of ground-living termites such as *reticulitermes*, that usually means fungal infections. What to do? Well, as in the case of many eusocial insects (colonies in which there is a division of labour), the termites get organized. During an outbreak, a group is organized, quite literally, to bury the dead. These undertakers are risking their own lives, but the colony is all.

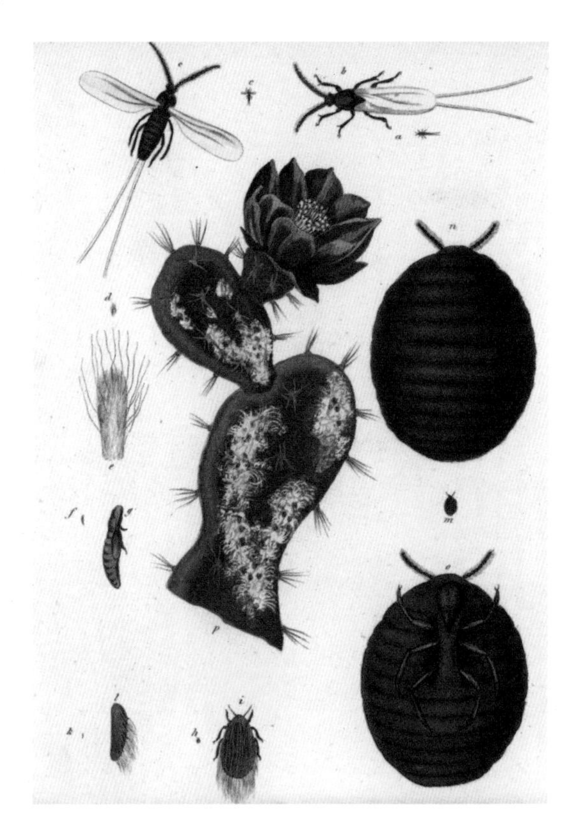

COCHINEAL

Dactylopius coccus **| Hemiptera / Dactylopiidae**

The pigment produced by this tiny insect has had a monumental effect on European art and culture, but only because Europeans had seen the Aztecs processing and using it with incredible results.

Red was a difficult colour to produce in Renaissance Europe. Most attempts to create a rich red pigment failed, and those that were successful were so expensive that they were only affordable to the richest. Unbeknownst to the 'enlightened' Europeans, across the Atlantic Ocean the Mesoamericans had, for almost 2,000 years, been producing a red pigment of eye-popping intensity, harvested from a tiny bug. *Dactylopius coccus* is a scale insect that lives in arid regions of Central America. The wingless females produce carminic acid, a crimson bio-defence which is processed into carmine – the deep, red dye that was prized by the Aztecs and used as currency in their marketplaces. White European settlers exploited the Aztecs and brought large quantities of the dye back to Europe, where it became the favoured pigment for Renaissance artists who used these startling hues to symbolize wealth, power and authority, and emphasize figures of religious importance in their paintings.

LORD HOWE ISLAND STICK INSECT

Dryococelus australis **| Phasmatodea / Phasmatidae**

The world's largest was once the world's rarest.

Lord Howe Island lies in the Tasman Sea, 600km (370 miles) from the east coast of Australia. It is a remote place, just 14 sq. km (5½ sq. miles) in extent. And like many such places, some strange fauna has evolved in isolation, none stranger than the world's heftiest stick insect, a jet-black, shiny 'tree lobster', 20cm long. In 1918 a shipwreck unleashed black rats on to Lord Howe Island and killed off all the stick insects. End of story? Not quite. Ball's Pyramid is a 600-m (2,000-ft) tall sea-stack that lies 20km (12½ miles) from the main island and in 2001, incredibly, 24 survivors were found under a single *Melaleuca howeana* tree in this inhospitable and unlikely place. A world population of 24 makes this the rarest insect ever recorded. Some animals were eventually removed and bred in captivity, and the population now stands at over 10,000. Rats are being eradicated from the main island and the stick insects, it is hoped, will soon thrive there again.

Common Carder Bumblebee

Bombus pascuorum **| Hymenoptera / Apidae**

The appearance of the first queen bumblebee is a delightful sign of spring.

Seeing the first bumblebee of spring is a joyous sight. In the UK, it might even happen on the occasional sunny day in late February. Bumblebee queens hibernate in leaf litter, behind bark, in the soil and in holes in trees. Before they enter hibernation in the autumn, they have already mated, so in early spring they are primed to start a colony.

To do this, they must first feed voraciously, so they visit every spring flower they can. A bumblebee at a crocus or a deadnettle, or among early blossom, is redolent of the busyness, energy and optimism of the season.

MUSCA

Insects are everywhere – even in the 'heavens'.

In the southern night sky, just south of Crux (the Southern Cross) sits the only constellation of an insect. Musca was mapped in the late 16th century and has been known officially as several translations of the word 'fly' over the centuries: De Vlieghe (Dutch), La Mouche (French) and Musca Australis (Latin for 'southern fly'), as well as a 17th-century sojourn as Apis (Latin for 'bee'), before being shortened to its present incarnation. The triangle shape, resembling the outline of a housefly, gives Musca its name. It is an arrangement of six stars; the brightest, Alpha Muscae, is a blue-white sub-dwarf star. Beta Muscae, a binary system, is 340 light years away, whereas the orange giant, Delta Muscae, at a distance of just 91 light years, is relatively close to our home star system. It also contains the bizarre and beautiful Spiral Planetary Nebula and the Engraved Hourglass Nebula. Unfortunately for Musca, it resides in the sky just within tongue's reach of its celestial neighbour, Chamaeleon, with which it is destined to remain locked in an eternal cosmic food web.

Indian Stick Insect

Trachythorax sparaxes

Phasmatodea / Lonchodidae

What are you doing for the next three months? For individuals of the Indian stick insect, the answer could be 'mating'. This animal holds the insect record for the longest time when male and female are joined together, which stands at 79 days.

Copulation itself is probably fairly infrequent. The male is there to be present when the female is receptive, and also there to prevent any other males inseminating her. So, he mounts her and just sticks around, if you'll pardon the pun. He is so small that he is probably no more inconvenient than a rucksack, and the female can get on with her life.

Ladybird Bug

Steganocerus multipunctatus **| Hemiptera / Scutelleridae**

Above: The ladybird bug does a terrific job of mimicry, by sharing the colour characteristics of a neighbouring beetle and spider.

Left: Male Indian stick insects are significantly smaller than their female counterparts.

Is it a bug? Is it a beetle? It is, in fact, a true bug – a kind of shieldbug, to be more specific. The ladybird bug is resident in sub-Saharan Africa, where it lives with very similar-looking species, *paraplectana thorntoni* – a small orb-weaving spider and a tortoise beetle. They have evolved to look like each other, an example of Mullerian mimicry, in which species adapt similar visual characteristics in order to improve their chances of survival. The ladybird bug also looks quite like a ladybird; a black ground colour with pale spots – also a good tactic as many ladybirds are distasteful or toxic. What sets this true bug apart from beetles is its rostrum; a long, rigid mouthpart which is held underneath it, instead of the mandibles which are typical of beetles.

Blue Morpho

***Morpho menelaus* | Lepidoptera / Nymphalidae**

The colourful scales of the blue morpho are partially waterproof.

Morphos are the iconic butterflies of the Amazon rainforest, although they occur throughout Central and South America. They could have been thought up by a PR company – huge, shining blue, showy and fairly common. Seeing one in the wild as it zooms past you in a clearing or along a forest trail is one of the definitive, shivers-down-the-spine experiences of the region.

Morphos drink the juices of overripe and rotting fruit and, when they land, they disappear from sight, bearing only modest brown colouration, and eyespots, on the underwings. Only the males sport the remarkable electric blue. The iridescence is not technically a colour, since these butterflies have no pigment, but the scales on the wings have nanostructures that break up light and only reflect blue wavelengths. Intriguingly, the scales also repel water, but in a proper tropical downpour these show-offs have to take shelter.

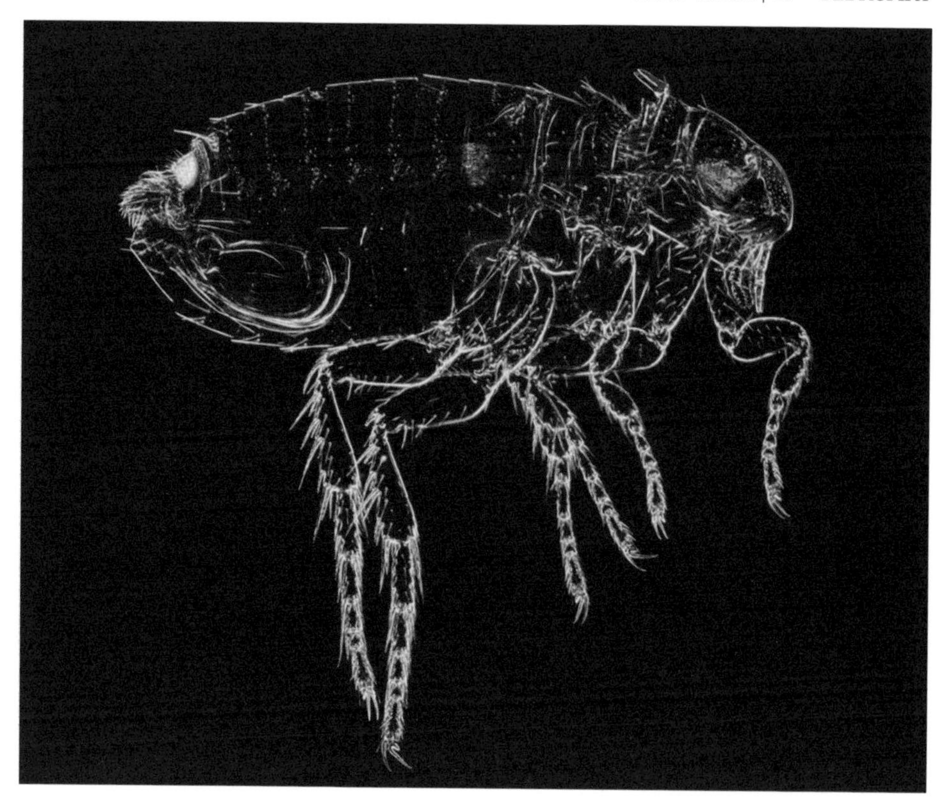

Domestic Fleas

***Ctenocephalides* spp. | Siphonaptera / Pulicidae**

No leap year would be complete without our best-known jumping insect, the flea.

Don't get all jumpy, but let's talk about the greatest leap of all. The domestic flea can jump the largest distance relative to its own body weight of any known animal; individuals have been recorded leaping up to 160 times their own body length. But the flea doesn't just go straight up and down like a pogo stick; it launches into a beautiful arc, ascending across an ecliptic which averages around 12cm high and 20cm in length. The reason they jump in this manner is all down to dispersal method. Fleas are wingless – they can't simply fly to another host, so instead they have evolved a self-propulsion system with a perfect trajectory to launch themselves onto any animals in close proximity. They have a 'pleural arch' on the thorax that bends and snaps back with explosive force (likened to the energy released by an archer firing a bow), catapulting them at a velocity of 1.9m per second, subjecting them to forces of up to 200G as they fly through the air to their new homes.

BARK LICE

***Neotrogla* spp. | Psocoptera / Prionoglarididae**

Female bark lice have an organ much like a penis.

The insect world is a marvellously diverse place containing every size, shape, colour – and gender. While we refer much to 'female' and 'male' in this book, this is purely from a biological standpoint. Due to their short lifespans, a lot of insect behaviour revolves around reproduction, so we refer to sexes in the binary form but in reality, sexual and gender fluidity is very common. Asexual reproduction is incredibly common, as many species have evolved to reproduce without the need for biological males. Gynandromorphy is often observed, where an insect is bilaterally divided into one half male, one half female. In this species of bark louse (*neotrogla*) females have a penis-like appendage and males have a receptive opening. There are no rules here, and we love it.

Shining Leaf Chafer

***Chrysina limbata* | Coleoptera / Scarabaeidae**

It's as brilliant as the sun glinting on a drop on a leaf – and that's the idea.

There are many lustrous shiny beetles in the world, but few can compete with Costa Rica's remarkable *chrysina limbata*, sporting surely the most perfectly polished silver finish of any insect. The species shimmers so much that scientists have exhaustively studied its elytra (wing casings), finding an extraordinarily complex multi-layered arrangement of chitin that reflects 97 per cent of all light. The surface has over 70 layers of chitin, each with different refractive indices, and light is both reflected and amplified.

But why be so shiny – doesn't it make a beetle obvious? No, in the wet forests where this beetle lives, droplets of water are forever on the leaves. The beetles resemble these droplets, being afforded the most exotic of cryptic camouflage.

GIANT WATER BUG

Lethocerus maximus **| Hemiptera / Belostomatidae**

Giant water bugs are superbly evolved for water locomotion, with powerful legs for swimming.

The giant water bug is the largest known aquatic insect in the world, its body length reaching in excess of 9cm (3½in) – big enough to fill the palm of your hand. Native to the tropical biomes of North and South America, it can be found in streams, lake margins and ponds. It lurks in vegetation below the surface, where it patiently lies in wait for prey. An ambush predator, it propels itself forwards rapidly through the water with its back two pairs of legs, grabbing hold of unsuspecting invertebrates with its huge, grappling forelegs. Male giant water bugs are proactive parents. A female lays her eggs on a male's back following mating, attaching them with a gluey substance; he will then look after the eggs until they hatch.

Pied Parasol

Neurothemis tullia **| Odonata / Libellulidae**

The pied parasol is a very common south Asian species.

This petite damselfly is so strongly associated with rice fields that it is sometimes called the pied paddy skimmer. Its distinctive black-and-white colouration, delicate undulating flight and small size, make it unmistakable. In common with many damselflies, an individual's horizons are quite limited. Mark-recapture experiments of this species in Malaysia found that both males and females had home ranges with a radius of 30m (100ft) and few strayed beyond 130m (425ft).

Namib Desert Beetle

***Onymacris spp.* | Coleoptera / Tenebrionidae**

Above: In the driest of climates, Namib desert beetles can seemingly 'magic' water from the air.

Opposite: The enormously long rostrum of this weevil is used as an exploratory probe and drill.

In one of the planet's most arid areas, a small beetle has evolved an extraordinary survival strategy. The Namib Desert in southern Africa has hardly any rainfall but, in the right temperature and humidity conditions, a slow, rolling fog descends across the tanned dunes to bring temporary relief to the dryness. *Onymacris* is one of many species of darkling beetle to inhabit the desert, but it is only one of two known species that will actively collect fog dew, which it achieves by scaling the dune and assuming a 'downward dog'-style yoga position, with its back at just the right angle to the moving fog blanket. Dew accumulates on the beetle's highly hydrophobic exoskeleton, and then gravity assists the droplets to run down specially adapted grooves on the elytra into its mouth, like rain along guttering. This unlikely pairing of behaviour and body plan works extremely well and provides welcome rehydration in one of the most hostile habitats on Earth.

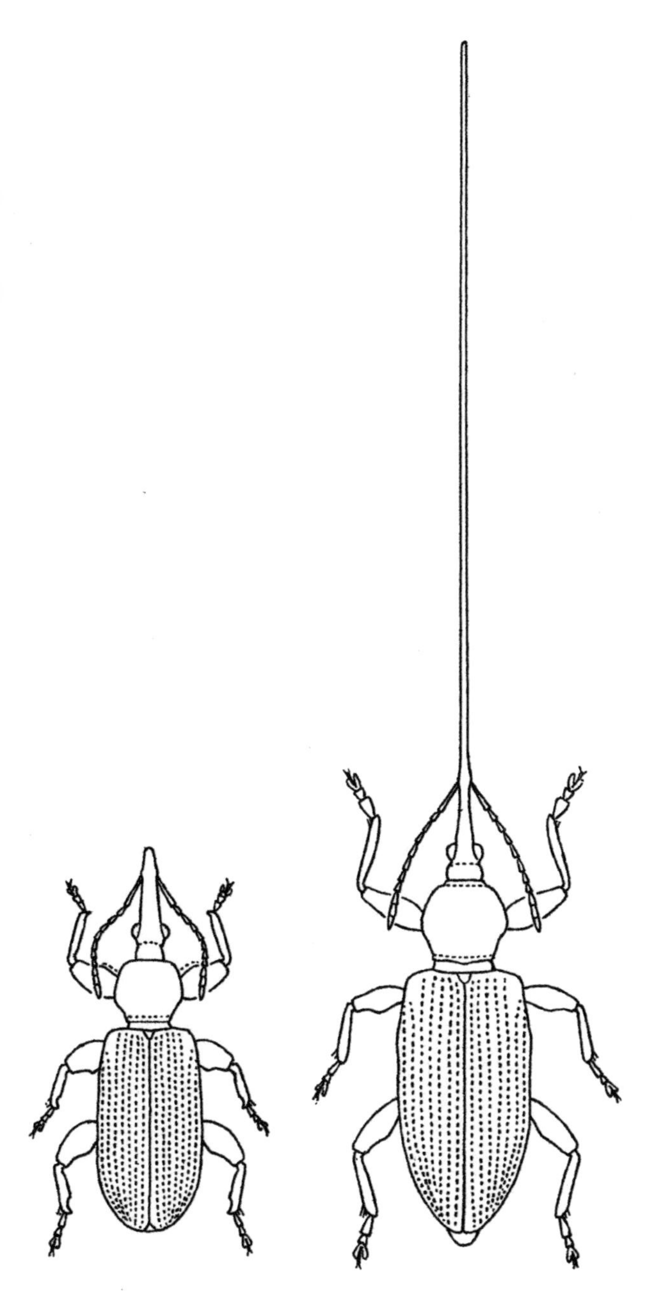

HOSE-NOSE CYCAD WEEVIL

Antliarhis zamiae | **Brentidae**

The world of weevils is a weird and wonderful one, particularly when it comes to the front end. Weevils are known for their fabulous rostra – the part of the face that we might identify as the nose. They can be short and stout, mildly elongated, long and slim or, in the case of this particular specimen, interminable. The female hose-nose cycad weevil of southern Africa has a splendid snout, at least twice the length of her body, which she probes into the cones of cycad plants and drills a hole through the surface of a seed deep within. She then lays an egg through the hole, and within these the larvae develop, eating the nutritious innards of the seed before pupating and emerging as adults. Males are easily distinguished from females, bearing a significantly shorter rostrum.

Question Mark

***Polygonia interrogationis* | Lepidoptera / Nymphalidae**

The question mark butterfly is a welcome sign of spring in temperate North America.

The snow is on the ground, but it's a sunny day in very early spring. Many a heart is gladdened by the appearance of the glorious question mark, one of the earliest butterflies to appear in temperate North America. The unusual name comes from a pale silvery patch on the underside of the hindwing, which looks vaguely like a question mark. A closely related species has a similar mark and is called a comma (*Polygonia c-album*).

The question mark hibernates as an adult and might appear in midwinter on an unseasonably warm day, but it can return to hibernation. This species can eat rotting fruit and sap if nectar isn't available.

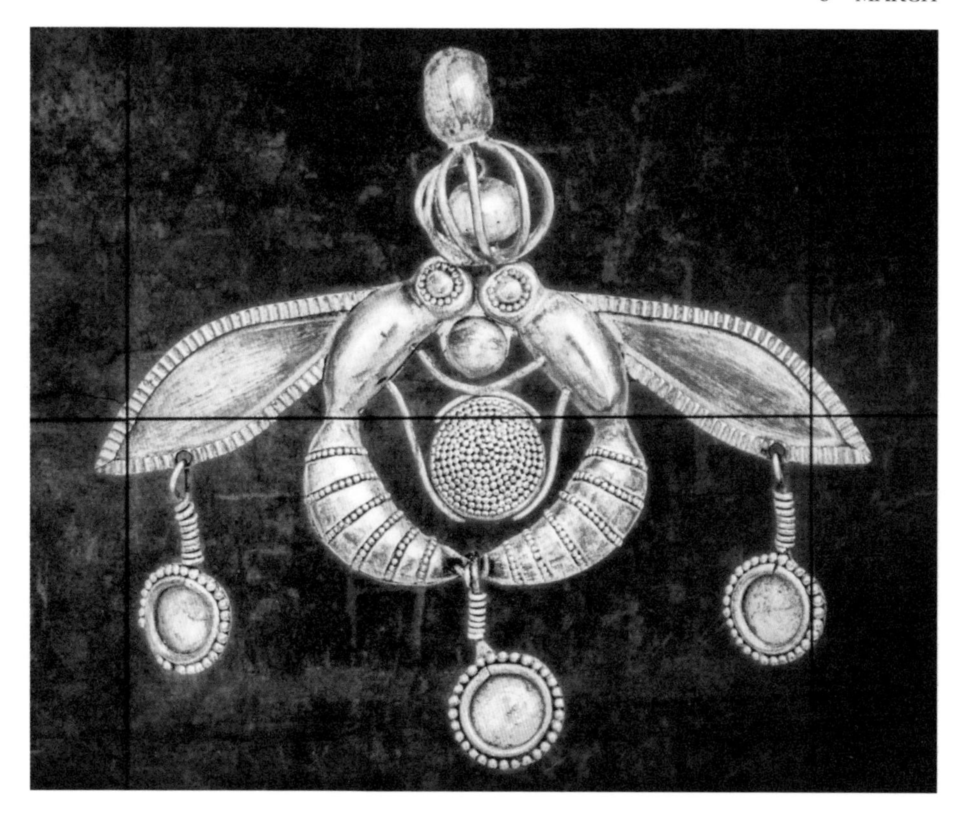

The Bees of Malia

This gold pendant of Minoan origin was excavated from a palace in Malia, Crete, and dates to around 1700 BCE.

The Heraklion Archaeological Museum in Crete is home to a very early representation of insects. The Bees of Malia, a pendant cast in gold of two hymenopterans, is Minoan and over 1,500 years old. There are various opinions of the identity of the insects; originally thought to be bees, some now believe they are hornets or potter wasps. Two individuals appear to rest on a round object, purported to be a honeycomb, mirroring each other's posture. They conjoin at the head and the tip of the abdomen, forming an outer circle, with their wings outstretched behind them, and appear to be depositing a droplet of nectar or honey. Beneath the 'bees' dangle three circles, which may portray the clay pot beehives used by the Minoans. Though the provenance and meaning of the pendant is a mystery, it's nice to think that the Minoans may have been paying tribute to the importance of honey bees in food production, as honey would have been invaluable as a sweetener, as well as having medicinal benefits.

Chequered Cuckoo Bee

Thyreus caeruleopunctatus | **Hymenoptera / Apidae**

Bright blue and black chequered cuckoo bees are unmistakeable as they zip around summer flowers.

Can there be a more curious sight in an Australian garden flower border than the powder-blue and black chequered cuckoo bee? This fabulously marked, flying fuzzball is quite the spectacle as it zips erratically around flowers, looking for the best nectar. It is superficially similar to the blue-banded bees, and for good reason, because the female chequered cuckoo is just that – a cuckoo that secretes its eggs into the brood chambers of blue-banded bees, to save itself the trouble of building its own nest. The young interlopers that hatch out steal all the provisions and emerge to cut their own marvellous swathe in the garden.

Brimstone Butterfly

***Gonepteryx rhamni* | Lepidoptera / Pieridae**

The bright, breezy brimstone is the epitome of spring in Eurasia.

This delightful, pale yellow butterfly is one of the signs of spring in Britain and Europe, routinely on the wing in March. Brimstones overwinter as adults, so despite their fresh appearance, these early season beauties are already several months old. A few persist into July, lasting almost a year in the adult stage, exceptionally long for any butterfly.

These lively butterflies always seem to be in a hurry, and they are certainly wide-ranging. The males patrol for females, while the latter make life hard for themselves by only laying eggs on two bushy plants, purging buckthorn and alder buckthorn, which grow at low densities. Furthermore, each female only lays a single egg in each bush.

Mirror Orchid Scoliid Wasp

Dasyscolia ciliata **| Hymenoptera / Scoliidae**

Above: Mirror orchids have evolved a cunning way to trick some male scoliid wasps into pollinating them.

Opposite top: Bluebottles get everywhere!

Opposite bottom: These bugs slide over the surface of the water.

Evolution is random, and it takes millions of years. In that time, completely unrelated organisms can establish the most remarkable relationships. *Dasyscolia ciliata* is a wasp that has had a remarkable effect on *ophrys speculum*, the mirror orchid, whose flower evolved to resemble the female wasp. The flower has a smooth, metallic blue patch that mimics shimmering, iridescent wings; the outer edge is hairy, replicating the wasp's fuzzy body. The reason this orchid looks like a female scoliid wasp is pollination; resembling a female wasp gets loads of attention from male wasps, which are fooled by the subterfuge and attempt to mate with the orchid flower. This trickery is pseudocopulation; the male swoops onto the flower and grabs it, thinking it is clutching a female from behind. The pattern mimicry is such that the male's head is perfectly positioned to brush up against the anthers, which coat him with pollen. He then flies off to try it on with other orchid flowers, transmitting the pollen as he goes.

Bluebottle

Calliphora spp.

Diptera / Calliphoridae

If there was ever a fly-on-the-wall documentary about an actual fly on the wall, it would be worth watching. How does a bluebottle perch effortlessly on vertical surfaces and walk on the ceiling? Well firstly, it has six legs, each with a holding claw at the tip. Secondly, at the base of each claw is a 'pulvillus,' like a pad fitted with a dense brush-tip of hairs. And finally, these hairs are coated with sticky fluids. Easy.

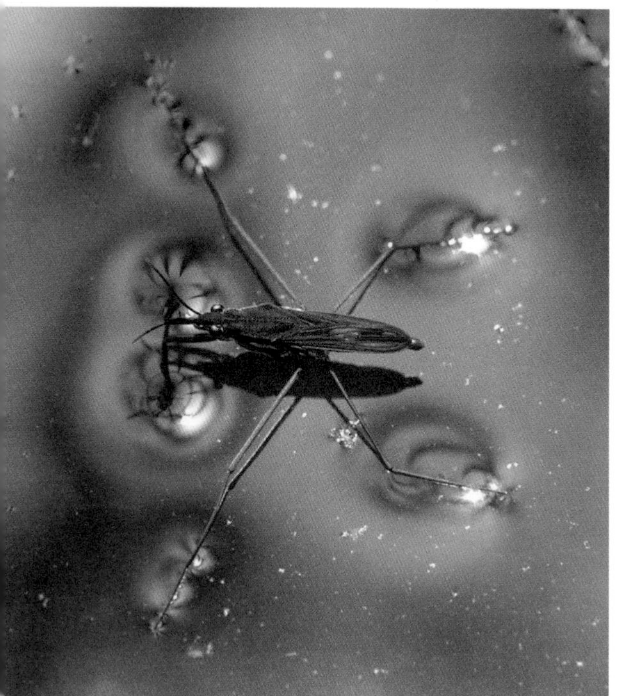

13TH MARCH

Common Pond Skater

Gerris lacustris

Hemiptera / Gerridae

It could be exhilarating to be a pond skater. This predatory bug lives on the water surface and 'rides' the surface tension, sliding at speeds of up to 1.5m (5ft) per second. It is very light, has a waxy cuticle and has long legs that spread widely. Not only do they walk on water, some pond skaters can also fly!

Orange Oakleaf

Kallima inachus | **Lepidoptera / Nymphalidae**

Blink and you'll miss it. And that's the point; this butterfly wants to be overlooked by predators.

The orange oakleaf butterfly of tropical Asia appears to have the best of both worlds. Open-winged, it is a blend of dazzling beauty and self-defence: bands of orange and blue iridescence and false eyes – typical of many butterflies – scream 'stay away!' at potential predators. Yet, once those astonishing wings are closed, the shape and colour of the underwing transform entirely, exactly matching that of a battered, dried, fallen leaf, and when resting motionless on the forest floor it becomes virtually imperceptible, perfectly reflecting why it is also commonly known as the dead leaf butterfly.

Hairy-footed Flower Bee

Anthophora plumipes | Hymenoptera / Apidae

This popular spring bee lives longer in cooler weather.

Have you ever wondered how long the insects in your garden live? It's a frequent question, although laborious to answer, since scientists studying wild populations must mark and recapture or resight individuals and make their projections based on negatives. Nonetheless, several studies of this delightful solitary bee, which is an early emerging species in Europe, suggest that males live for about 3–4 weeks, and females for 5–7 weeks. However, they live longer in cooler weather, with their activity being less intense. The occasional individual manages to reach the grand old age of 10 weeks.

The Largest Insect We Know Of

Meganeuridae

Just imagine how amazing it would be if modern-day dragonflies were as large as crows.

As far as we currently know, the largest insects to ever inhabit the Earth were an ancestor of our modern-day dragonflies. A fossil was discovered in Kansas, USA, in 1939, proving the existence of a species with a wingspan of around 70cm (over 2ft) and a body length of 43cm (17in). Named *meganeuropsis permiana*, it was present during the Permian period, 200–300 million years ago, well before birds and mammals existed. If these beasts were still around now, you wouldn't miss them; they would have a similar wingspan to a crow and, with their aerial hunting prowess, would probably still rule the skies. The late Permian mass extinction event ended the reign of Meganisoptera, and it is thought that factors such as the lower atmospheric oxygen level and reduced prey availability in the aftermath meant that these leviathans were increasingly unable to sustain their own body size, bringing to an end the reign of the giant flying insects.

Giant Mesquite Bug

***Thasus neocalifornicus* | Hemiptera / Coreidae**

The giant mesquite bug doesn't have a 'drab' phase – it looks spectacular at every stage of its life.

For those insects that don't go through a pupal stage, life is one long catwalk, filled with eye-popping wardrobe changes; from egg to adult, they go through a number of external transformations. Because their tough, rigid exoskeleton does not stretch as they grow, instead they shed this outer layer entirely for a new one with a little more room; a process that can be undertaken several times. Each growth phase – called an instar – gives the bug a different appearance to the previous, and this is thought to improve the bug's chances of survival by becoming unfamiliar to its known predators. Hiding in plain sight, as it were. The giant mesquite bug has a fabulous repertoire. Through its five instars it develops thick red bands on its black legs, only to lose them again in adulthood. Its small red abdomen develops a ribcage-like pattern and it begins to resemble a power lifter in the lower leg department. But it's not only its appearance that changes. Research has discovered that the chemical pheromones emitted by the giant mesquite bug also change from nymph to adult, another adaptation that can protect the bug by masking its true identity.

Dark-edged Bee-fly

Bombylius major **| Diptera / Bombyliidae**

Bee-flies are tiny flying teddy bears with a fascinating dark side.

Is it a bee? Is it a fly? Who knows? Well, we do really; it is a fly that uses its disguise in a way that is ingenious, and just a little bit gruesome. The dark-edged bee-fly is a fuzzy fluffball that is more bear than fly. Large bronze eyes and a long, rigid snout give it a decidedly inquisitive expression. It zips and hovers around flowers like a tiny drone, hovering in mid-air to insert that extensive proboscis into spring flowers such as lungworts and primrose – this flying ball of candyfloss is just the sweetest little floofball, but it has a heck of a kick on it. The dark-edged bee-fly, for all its sweet exterior, is a kleptoparasite, meaning it lays its eggs in the nest burrows of solitary bees in the most incredible manner, by flicking them from its rear end, using its hind legs to drop kick the weaponized ova into the burrow entrance. The larvae then feast on the contents of the burrow, eggs, provisions and all before emerging as a fluffy bear to fool us once more into thinking it is entirely sweet and harmless.

O'AHU DAMSELFLY

Megalagrion oahuense **| Odonata / Coenagrionidae**

The glorious O'ahu damselfly has an unusual larval stage.

When living things are isolated for a long time on islands, they often evolve in highly individual and unexpected ways. Take the O'ahu damselfly, known locally as *pinapinao*. It is a lovely pink-red damselfly that does everything in a damselfly-ish manner: flutters gently about, rests with wings folded behind its back, and so on. But the larva is a total outlier – in an outlier archipelago. It is the only damselfly in which the larvae are terrestrial, living in leaf litter instead of the usual fresh water.

Insect Idioms

Insects are woven into the fabric of our language; we refer to them often without even realizing it.

Even if you don't like insects, chances are that you talk about them more than you realize. There are many idioms related to insect appearance and behaviour – some complimentary, some less so. We've all wanted to be a fly on the wall, and we've certainly all been as busy as a bee. When faced with a challenge, we feel butterflies in our tummy; and ever had ants in your pants, when you just can't sit still? 'Por si las moscas' is a Spanish saying for wanting to avoid an undesired consequence; translating roughly as 'in case of flies' and said to derive from the act of covering food to prevent flies landing on it.

Here are some other insecty idioms that form part of our everyday language: The bee's knees – Social butterfly – Snug as a bug in a rug – Dropping like flies – Make a beeline – Sleep tight, don't let the bedbugs bite – Creating a buzz – Hive of activity – A bee in your bonnet – Wouldn't hurt a fly – Beetle about – Fleabag – Bugbear – Bugging me – Fly in the ointment.

Fruit Fly

Drosophila bifurca **| Diptera / Drosophilidae**

Did you know that the fruit fly is one of the world's most studied organisms?

As mentioned elsewhere in this book, the fruit fly (*drosophila melanogaster)* is one of the world's most intensely studied organisms, especially in the field of genetics. It is far and away the best known of the genus *drosophila*, which contains more than 1,500 species around the world.

Let's hear it, then, for *D. bifurca*, which has recently (1995) come out of its relative's shadow. Why? Because the male has the longest sperm relative to body size of any animal in the world. The sperm is 58mm long, which is 20 times longer than the fly itself. The testes are 67mm long and constitute 11 per cent of the animal's mass.

22ND MARCH

Winter Cherry Bug

Acanthocoris sordidus

Hemiptera / Coreidae

If you've got six legs, you might as well use them to the max. The winter cherry bug has its back pair modified for combat, sturdy and fitted with spines. When a territory-holding male is challenged by a rival, the two combatants turn their backs on each other, splay their legs and touch them. Each now tries to wrap its legs around the other's abdomen in a vice-like grip, a contest of skill and strength that can go on for many minutes.

23RD MARCH

Australian Hangingfly

Harpobittacus australis

Mecoptera / Bittacidae

These impressive insects are very unusual because they catch their prey with their hind legs, which have a broad span and are high movable. The name hangingfly comes from their habit of holding on to a stem with their front limbs and hanging down like a trapeze artist. The hind limbs grab, and the middle limbs help to manipulate the prey up towards the jaws, while they remain hanging. Prey may be caught in a stationary position or in flight and is overpowered by digestive enzymes.

Hurt No Living Thing

Above: Christina Rossetti felt a strong connection with nature.

Opposite top: Winter cherry bugs fight each other for high stakes.

Opposite bottom: The well-named hangingflies are voracious predators.

Despite spending the majority of her life in industrial, Victorian London, the Pre-Raphaelite poet and thinker, Christina Rossetti, was deeply fascinated by nature. She valued all life and took great pleasure in even the smallest details, highlighted here in a beautifully eco-centric poem that she wrote in 1872: 'Hurt no living thing'.

Ladybird, nor butterfly,
Nor moth with dusty wing,
Nor cricket chirping cheerily,
Nor grasshopper so light of leap,
Nor dancing gnat, nor beetle fat,
Nor harmless worms that creep.

Christina Rossetti (1830–1894)

Giant Edible Cricket

Tarbinskellius portentosus **| Orthoptera / Gryllidae**

Insects could be the future of food production.

Have you ever eaten a cricket? If not, the chances are that you soon will, or at least your descendants will. Consuming insects is the future; they are tasty and far more environmentally friendly to produce than most other meats. And it's hardly a new idea; people have been eating insects for 7,000 years. The giant edible cricket is among 60 species of crickets (and 1,900 species of insects) that are eaten worldwide, and conveniently lives in burrows on the edges of fields. It can be dug or flooded out, or it can be attracted to light. In markets in Chiang Mai, Thailand, crickets are skewered four to a bamboo stick, then fried in oil while you wait.

Blue Shieldbug

Zicrona caerulea | **Hemiptera / Pentatomidae**

The blue shieldbug changes colour radically during its different life stages.

When is a blue shieldbug not a blue shieldbug? When it's a nymph, of course. This shieldbug goes through a dramatic change in colour in appearance between life stages, a transition known as ontogenetic variation. Surprisingly, the blue shieldbug starts out life red; its small black body has bright scarlet markings on its abdomen. During its final moult it completely changes its palette into a striking metallic turquoise. Why? Well, ontogenetic variation is still subject to discussion, but it's likely that nymphs and adults have quite different requirements for success. The red colouration of the nymph is a typical aposematic feature; a signal to predators of distastefulness or toxicity. The adult, however, displays a remarkably similar colouration to its primary prey – leaf beetles (*altica* sp.). When you think about it, it all makes sense; blending in with your food is a highly effective way to hide in plain sight, and there's no chance of eating the kids by accident...

Teddy Bear Bee

Amegilla bombiformis **| Hymenoptera / Apidae**

This superbly fuzzy insect looks like a bumblebee, but it is actually a flower bee.

Bees are well known and well loved for their fuzziness. There is often a certain bit of a bee that is fuzzier than the rest, for instance, the body may be fluffy and the legs relatively hairless, or the head may display an impressive mane. And then there's the teddy bear bee, which has sprouted hair from every available pore, the result of which is more bichon frise than bee. This stunning pollinator has a dense, all-over golden pile that is quite short, and so neatly clipped that it looks as though it has just stepped out of the salon. The legs are also fuzzy, particularly the pollen baskets, which can stow a large quantity of pollen. Despite its hairiness making it look like a bumblebee, it is actually a flower bee, one of several species with dazzling eye patterns and broad abdominal bands that provide that extra zing of colour in the hot summer borders.

Gleaming Fruit Chafer

Chlorocala africana **| Coleoptera / Scarabaeidae**

The gleaming green version is one of many colour variants.

Out in the wild this brilliantly coloured beetle lives in African bush country and drinks the sap of a wide range of trees and shrubs, including *proteas* in South Africa. It has a wide distribution and exhibits an astonishing range of colours from one region to the next – some orange, some red, some green and some purple. They look so different that many have been described as separate species. This beetle has acquired a second, somewhat unwelcome career, as a popular pet that breeds quickly in captivity.

ANT

Hymenoptera / Formicidae

According to the Qur'an, Solomon was able to talk with animals, including ants.

Eid al-Fitr marks the end of Ramadan, a month-long period of fasting, followed by two days of festivities, in which Muslims around the world celebrate the prophet Muhammad. Another prophet, Solomon (Suleiman in Arabic), is said to have been able to talk with animals. The Qur'an tells of a meeting between Solomon and the ants:

Until, when they came upon the valley of the ants, an ant said, 'O ants, enter your dwellings that you not be crushed by Solomon and his soldiers while they perceive not.'
QUR'AN, 27: 18-19

MANTIS

Don't mess with a mantis, especially if you want to keep your head...

She stills herself, a green meditation,
angled with desire for aphid, moth.
Icon, on guard, she is threat posed as prophet.
A body of tricks, mischief made leaf,
flowering to thorn; a small violence.
Trauma is feast. Mantis, wild queen, her face is geometry at play;
a compass for the dead.

ELLA DUFFY

Cave Cockroaches

Blaberus spp. | Hemiptera / Blaberidae

These cockroaches are great at timekeeping!

The giant cave cockroaches of the Tropics are a fascinating bunch. They dwell in caves, where they feed on the food waste and guano (poo) of resident bats. Successful populations can grow abundantly, and individuals can live for up to two years. They are nocturnal, but what is incredible is that they can 'tell the time' in the perma-twilight. Even with the lack of daylight, these cockroaches manage to sustain a consistent circadian rhythm. They have an internal master body clock which helps them biologically discern night from day, and they also take cues from tiny changes in light levels, temperature fluctuations, wind disturbance and even the behavioural rhythms of the bats.

Resin Assassin Bugs

Gorareduvius spp. | Hemiptera / Reduviidae

Assassin bugs are among a tiny minority of insects that use tools. This is a tree resin assassin bug.

Assassin bugs are as dangerous as they sound. They run fast and ambush a wide variety of small invertebrates, then insert their sharp rostrum and suck the prey dry. In Australia, a few of these bugs have hit upon a very unusual way to improve their hunting success. They live among spinifex grass, which exudes a sticky resin. Before going hunting, they scrape the resin off the leaves and apply it to their bodies. It seems that, when covered in resin, they acquire just a little more adhesion on their prey, giving it less chance of escape. Scientists carrying out tests, which involved scraping resin off some individuals, found that the resin improved their capture rate by 26 per cent. It's a rare case of tool-using.

Carolina Sphinx Moth

***Manduca sexta* | Lepidoptera / Sphingidae**

The Carolina sphinx has been observed by scientists and artists alike for centuries.

One of the most familiar images in American entomological pop culture must be the tobacco hornworm. This gorgeous, green caterpillar, silky smooth to the touch and beautifully accessorized with jazzy go-faster stripes and that eponymous 'horn' at the rear. It does, however, have a soft spot for tobacco and tomato plants, which has gained it a less-than-sparkling reputation among commercial and domestic growers of these plants. When naturalist Maria Sibylla Merian made her botanical study of the life cycle of the tobacco hornworm, tomatoes and tobacco were a relatively recent arrival in Europe. Merian was born on the 2nd of April 1647 in Germany, into a prosperous printing business. We could say that she broke the glass ceiling for women in natural history; Merian was studying and recording the ecology of insects, decades before Carl Linnaeus began putting names to them. While pursuing her love of entomology, she gave birth to two daughters, rescued her mother from a terrible marriage, and left her own marriage to join a commune with her children, eventually divorcing her husband – all pretty kickass behaviour for a 17th-century woman. She travelled from Europe to Suriname with one of her daughters, where they studied and recorded a large amount of flora and associated

insect fauna. Pioneering as she was, we must acknowledge that Merian's work in Suriname directly benefitted from the slave trade, slave routes and exploitation of humans, as did most (if not all) natural historians operating in those areas at the time. She received funding from the Dutch government, themselves deeply embedded in the Transatlantic slave trade, and Merian herself received help with her research directly from local people who were enslaved. She observed and commented on the conditions in which the enslaved workers at the sugar plantations were kept, the trauma it caused them and how they used the peacock flower for the most tragic purposes. This quote by Merian accompanies the illustration in her folio, *Insects of Suriname* (1726 edition): 'The Indians, who are not treated well by their Dutch masters, use the seeds to abort their children, so that they will not become slaves like themselves .... In fact, they sometimes take their own lives because they are treated so badly, and because they believe they will be born again, free and living in their own land. They told me this themselves.'

Seven-spot Ladybird

Coccinella septempunctata | Coleoptera / Coccinellidae

Seven-spot ladybirds are the quintessential gardener's friend.

The seven-spot ladybird is quite possibly the most recognizable, and certainly one of the best-loved bugs in the British Isles. It has inspired poems, songs and folklore; ask children to draw an insect and this is invariably the first one that will appear on their paper. It has a reputation as the 'gardener's friend', due to the larva's propensity to hoover up aphids from prized garden plants. This is actually a task performed by many garden predators, though seven-spots seem to be given the lion's share of the credit. In recent years, the seven-spot ladybird has had to share its niche with the larger, competitive harlequin ladybird; the latter eats the same diet and will even eat other ladybird larvae. You can almost guarantee seeing the seven-spot all year round. Spring and summer see it most active as it mates and reproduces; in winter you can find them hibernating in crevices, dead plant stems and seed heads, and even in quiet nooks in your own home.

Red Swampdragon

Agrionoptera insignis | Odonata / Libellulidae

Dragonflies are unsurpassed aerial predators.

Always well turned out, the red swampdragon is a common dragonfly found in a broad span from eastern Australia all the way to Japan, where it occurs in streams, swamps and shaded pools. Here it spends much of its time chasing smaller insects in raptorial fashion. Dragonflies are the supreme aerial predators. All four wings are powered independently, giving them unsurpassed manoeuvrability, including flying upside down and backwards! Their compound eyes have up to 30,000 eyelets, and apart from seeing the colours that we see, dragonflies can also detect the ultraviolet spectrum. It's just as well they aren't a lot bigger, or we would all be in trouble.

Tarantula Hawk

Pepsis grossa **| Hymenoptera / Pompilidae**

This wasp packs a punch – its venom can fell a spider far bigger than itself.

Of the 15 species of tarantula hawk wasp found in the USA, *pepsis grossa* is one of the largest. The adults can be found nectaring on flowers in desert scrubland, but it will usually be the males that are observed, because the females are busy elsewhere, hunting tarantulas. She will ambush and paralyze an unsuspecting tarantula by quickly stinging it before she herself can be bitten. The immobilized tarantula is then taken back to the wasp's nest where it will become a meal for one of her larvae. Her venom is one of the most painful known to science; it sits high up on the Schmidt Pain Index and is described as: 'Blinding, fierce, shockingly electric.' Despite this terrifying description, and its ability to dispatch prey significantly larger than itself, the tarantula hawk is not dangerous or aggressive, and will not attack unless it feels threatened. It is also rather wonderfully the official insect of New Mexico.

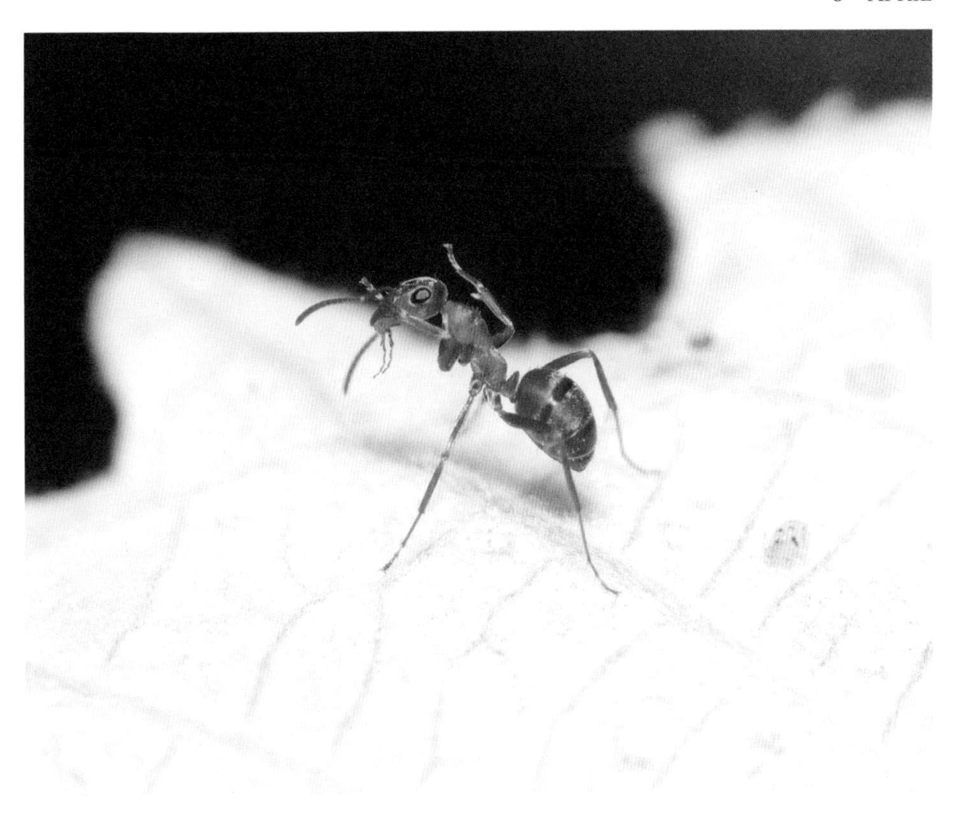

Silky Ant

Formica fusca **| Hymenoptera / Formicidae**

Ants have many extraordinary abilities – even self-medication!

Ant colonies are large, very closely knit gatherings of large numbers of very closely related individuals. Not surprisingly, one of the problems that they face is disease, which can be passed very rapidly through a colony. Fungal problems are among their worst. Recent studies have shown, though, that silky ants have developed a remarkable defence. When individuals are at the worst stage of infection, they suddenly change their diet to a type of aphid called *Megoura viciae*. Normally, ants milk aphids for their honeydew excretions, but in this case they eat the bugs instead. It's been shown that the aphids' tissues contain hydrogen peroxide, an antiseptic that kills fungal infections. Once they recover, they switch back to their normal diet.

LARCH LADYBIRD

Aphidecta obliterata | Coleoptera / Coccinellidae

Look closely at the branch of a larch tree, and you may spot one of these ladybirds.

Not all ladybirds have spots. Some are fairly plain-looking, which may work to one's advantage when attempting to blend into one's surroundings. The larch ladybird is a fairly uniform tan colour, with some minor deviations; the elytral suture (where the wing casings meet) is often darkened, and there are usually dark 'dashes' along the hind margins of the elytra. Native to coniferous habitats in Europe and North Asia, it is also now present in parts of North America, where it has been deployed as biological control for the accidentally introduced balsam woolly adelgid – an aphid-like insect that causes significant damage to fir trees. A single larch ladybird larva can consume over 2,000 adelgids, and it continues its prolific eating quest as an adult – quite something for a tiny, brown beetle.

SPECKLED WOOD

Pararge aegeria | **Lepidoptera / Nymphalidae**

Speckled wood butterflies are highly territorial.

Imagine that your world is a sunbeam. That's the case if you're a male speckled wood on its territory. These butterflies defend sunny spots within the shade of woodland. They spend much time basking, but if an intruder comes to challenge them, the two rivals embark on an aerial battle, flying against each other like two boxers locked together, rising ever upwards within the shaft of sunlight. Usually, the territory owner wins and the trespasser retreats, but the battles can last anywhere between 2 and 90 minutes. It's a treat to watch.

Some males don't bother to hold territory, but patrol the edge of the wood instead, hoping to come across a female.

Devil's Coach-horse

***Ocypus olens* | Coleoptera / Staphylinidae**

The devil's coach-horse is a fearsome hunter, and looks spectacular when it curls its abdomen over defensively.

Sometimes an insect develops a reputation that, over the years, reaches legendary proportions. This is certainly the case with the devil's coach-horse in Ireland, where it is known as the dearg-a-daol, steeped in superstition and even thought to embody the devil himself. The devil's coach-horse is a truly beautiful example of the huge family of rove beetles (Staphylinidae), most of which are fast, agile, ground hunters. It is a slender but robust beetle with a uniform matt black body. The elytra are tiny, not reaching further than the end of the thorax, and under these are amber wings that fold inside them by way of an astonishingly sophisticated hinge-lock system. The abdomen is completely exposed, and it is this that has become the devil's coach-horse's defining feature; when threatened it will curl the abdomen up and over its back, not unlike a scorpion. It's all show though, as this beetle does not sting, but it looks formidable enough to stop assailants in their tracks.

Marmalade Hoverfly

Episyrphus balteatus **| Diptera / Syrphidae**

Marmalade hoverfly population numbers can reach trillions as they gather together to migrate across Europe.

The marmalade hoverfly is a small fly that plays a huge role in our ecosystem as pollinator and 'pest' manager. The UK has its own small, resident population, but the majority of summer marmalades travel a couple of thousand kilometres from central Europe. They are one of several hoverfly species that undertake huge migrations across the continent every year. Up to 4 billion adults arrive in the southern British Isles between May and September, making them one of our most prolific pollinators, and the fact that they travel such long distances only increases their pollination potential further as they make pitstops through Europe, laden with pollen grains. But if the adult marmalade hoverfly's job is to create life, its progeny is preoccupied with ending it. The alien-like larva may look slow and a bit useless, but don't be fooled; it is a clinical and highly effective killer, hoovering up any aphids in its path.

11TH APRIL

Large-headed Resin Bee

Heriades truncorum

Hymenoptera / Megachilidae

Not all bees are big, fluffy and conspicuous. Many of the bees in our gardens and outdoor spaces are very small, and could be mistaken for flies. This large-headed resin bee is less than 10mm in length, and looks tiny when nectaring alongside its gargantuan bumblebee relatives.

12TH APRIL

Japanese Water Scavenger Beetle

Regimbartia attenuata

Coleoptera / Hydrophilidae

It's not unusual that bits of insects get pooped out by larger animals undigested, but some lucky individuals can actually survive intact. What is unusual about this beetle is the sheer speed with which it can get from one end to the other. Scientists have observed the Japanese water scavenger beetle exit in as little as 6 hours from entry, alive and well.

Above: Green-headed ants pass on chemical 'feedback' to each other.

Opposite top: A large head on a tiny bee.

Opposite bottom: Not many insects make it out alive once they are in the jaws of predators.

Green-headed Ant

Rhytidoponera metallica | **Hymenoptera / Formicidae**

In common with other eusocial insects, ants have a division of labour within the colony, with some members searching outside for food, others looking after the nest, and so on. Different colony members also have different food requirements: workers mainly consume carbohydrates, while the larvae need protein. Research on green-headed ants has shown that the larvae are able to provide feedback to the foragers, instructing them to make subtle adjustments to their efforts. Obviously, every ant about to go out looking for food says something like 'Your opinion is important to us. Please give us feedback by taking part in this survey...'

Bagworms

Lepidoptera / Psychidae

Bagworm moths have been reusing, recycling and upcycling far longer than us humans.

Upcycling is the name of the game with the bagworm, which is not a worm, but a moth. When they pupate, most moth caterpillars spin a cocoon from their own silk; however, the bagworm has a very creative approach. The larva collects together bits of leaf, twig, even sand and soil peds, then carefully builds the pieces up around itself, bonding them together with its own silk to create a cosy burrito in which to pupate. Female bagworms are usually wingless and do not leave their case; winged males have to locate them, although this is sometimes a futile process, as some females are parthenogenetic – a handy reproductive short cut when you can't travel far very quickly. Another successful strategy is to be eaten; females can be consumed by predators before they've even laid their eggs, but the eggs are hardened, and are excreted, which handily disperses them into new areas.

Lesser Water Boatman

Micronecta scholtzi **| Hemiptera / Micronectidae**

Water boatmen are bugs with a record-breaking talent.

What is the loudest sound made by any insect relative to its size? Almost everybody would say a cicada, or perhaps a cricket. But the record-holder, remarkably enough, is an aquatic bug, a lesser water boatman. The sound is so loud that people with good hearing can detect it by the waterside. When measured underwater, the song of this male water boatman registered 79 decibels. To give you an idea, coming from such a small animal, that's equivalent to a passing freight train.

Mantisflies

Neuroptera / Mantispidae

Most mimics are harmless versions of the animal they copy, but a mantisfly is every bit as dangerous within the microcosm as a true mantis.

Insect diversity is a delicious spectrum of morphology (the way things look). Evolution has played with every conceivable permutation of wing, leg, antennae, body shape and pattern to the point where it would be hard to conceive coming up with anything original. The only thing possibly more exciting than this is where morphology overlaps: whether for the purposes of mimicry or sheer coincidence, completely different insects that look like each other are a joy to behold, so behold, indeed, the mantisfly. It's a fly that looks like a mantis, until you look at it closely and spot the smoking guns – those little anatomical details that give away its true identity, such as the clear, veined wings, bulky abdomen and shorter, stockier antennae. But those front legs are an incredible bit of subterfuge that would make the most seasoned entomologist look twice. Mantisflies have huge, modified front legs just like those of mantids, and they use them in the same way, to grab and ensnare prey at lightning speed.

Slave-making Ant

Polyergus rufescens **| Hymenoptera / Formicidae**

Most ants work hard, but not all of them...

Ants have a well-earned reputation for hard work, but some, surprisingly, are lazy. Take the slave-making ant. This species gets a host ant to do all its hard work – to find food, look after the nest and raise and feed the young. It all begins when a young, newly mated slave-making queen needs to join a colony. She enters the compass of a different species, often *Formica cunicularia*, by tiptoeing into the functioning colony and emitting a pheromone to reduce the workers' aggression. Once inside, she kills the incumbent queen and in the changed circumstances the host members feed and look after her. She starts laying her own eggs, which then grow alongside the eggs of the host species. The host workers tend the larvae of both species and the host larvae grow up to perform all the colony tasks. The slave-making offspring, however, have only one task: to raid nearby nests of *Formica cunicularia* and bring the pupae and larvae back to join the slave hordes.

Soldier Fly

Diptera / Stratiomyidae

The broad centurion is a colourful, metallic hoverfly found in grassland.

Flies aren't the most popular insects, but here's a family that may change your mind. The soldierflies are a group of wonderfully coloured and patterned flies, many of which are associated with wetlands as they have aquatic or amphibious larvae. This one, the metallic-looking Broad Centurion (*chloromyia formosa*), is common throughout Europe, and one of the species most likely to end up in your garden. Like many flies, male soldierflies have large, holoptic eyes that meet in the middle of the head, distinguishing them from females.

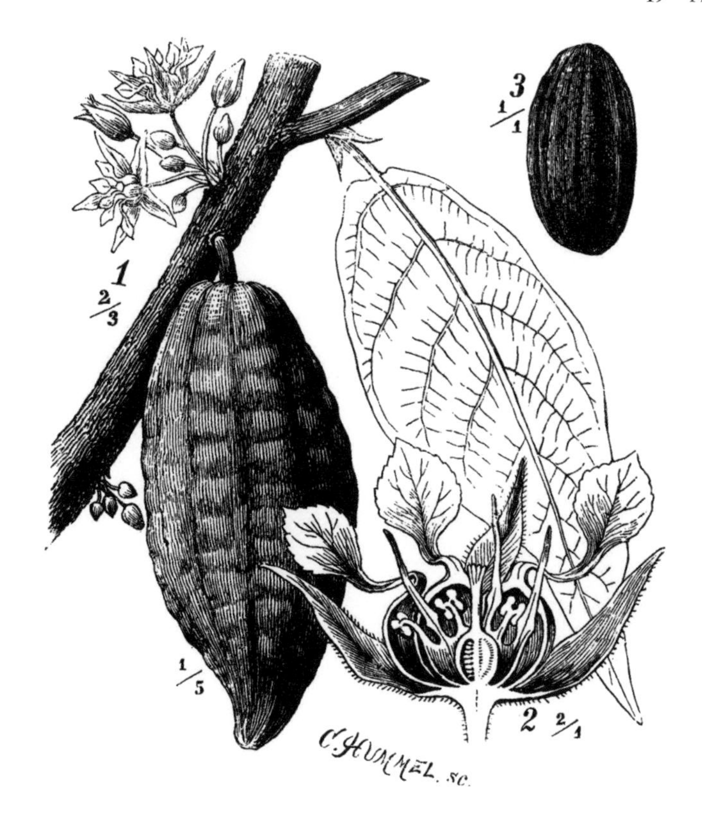

The Real Easter Bunny

***Forcipomyia* spp. | Diptera / Ceratopogonidae**

Many of us can't imagine a world without chocolate, and it's all thanks to this unsung hero, not a rabbit.

At Easter, the thoughts of many a sweet tooth – young and old – may be dominated by the thought of tucking into chocolate eggs, which are traditionally delivered by the Easter bunny in many parts of the world. But let us, for a moment, turn our attention to the animal that really does ensure we receive our festive chocolate – the cacao midge, a small fly that is responsible for the pollination of the flowers of *theobroma cacao*, the plant that produces cacao beans. Due to their size, *forcipomyia* flies are almost solely responsible for all global chocolate production. At just a few millimetres long, they can fit through tiny gaps in the cacao flowers to access nectar. Pollen grains stick to the flies, fertilizing the flowers so they produce fruit. And, my goodness, do we need these midges; as little as one in five hundred cacao pods will mature into yummy chocolatey potential, so let's hear it for the unsung heroes of our Easter treats.

Painted Snout Bug

***Eddara euchroma* | Hemiptera / Fulgoridae**

Every corner of the world has insects to take the breath away.

People often ask: 'What makes you love insects so much?' Well, perhaps this picture of the painted snout bug will help to explain part of their appeal.

Ask someone in South Africa if they've heard of it, and they almost certainly won't have. It is barely known; scientists aren't even certain exactly what it eats. Yet this totally obscure minibeast is beyond stunning.

Every corner of the world hosts insects as beautiful as this. They are the greatest set of hidden gems there is. They are the epitome of extravagant abundance.

False Ladybird

Endomychus coccineus **| Coleoptera / Endomychidae**

It looks like a ladybird, but this is actually a fungus-eating beetle in ladybird cosplay.

In spring and summer when you are out for a wander in the woods, have a close look at the bark of trees, especially those populated with fungus, because you may catch a glimpse of a most delightful little beetle. But look closely, because, as its name suggests, you may overlook it, assuming it is instead a ladybird. The false ladybird is rather smaller than our most common red and black ladybird species – almost half the size – as well as being flatter and more pear-shaped. Its bright, cheerful, red body with decisive black markings (more oblong than round) do give it a decidedly ladybird-like demeanour. This beetle does not eat aphids, though; its diet is exclusively fungus, more specifically fungi that grow on and in the bark of trees. Its aposematic colouring should be enough to deter predators, but if they persist, the false ladybird has a backup plan: it exudes foul-tasting liquid from joints in its legs, allowing it to escape while its attacker is left reeling with a mouthful of ickiness.

Horn-faced Bee

***Osmia cornifrons* | Hymenoptera / Apidae (Megachilidae)**

This pollen-dusted bee is living up to its reputation.

This is a great time in the Northern Hemisphere to appreciate blossom. There are fewer more glorious sights in nature. And to what do we owe this delight? Well, pollinators of course. No pollinators, no flowers.

If there is a title for the world's most effective pollinator, how about the horn-faced bee? Ask the Japanese. They introduced this bee from elsewhere in Asia in the 1940s, and now it is used in most of their apple orchards; it was further introduced into the USA in the 1970s. It requires 750 horn-faced bees to pollinate 1ha (2½ acres) of orchard; the same task would require 50,000 western honeybee (see page 25) workers. A single horn-faced bee can visit 2,500 flowers a day.

South American Dead Leaf Mantis

Acanthops falcata | **Mantodea / Acanthopidae**

The dead leaf mantises are so well camouflaged that they struggle to find each other!

One can only marvel at the extraordinary camouflage of the many mantis species that look like dead leaves. In this species, the male is small and brown and its long, functional wings look like a rolled leaf, while the flightless female looks like a curled and crinkled leaf. Both are almost impossible to see in the wild.

This creates a problem. The species' camouflage is so good that the mantids themselves can't find each other either. So the female has a contingency plan. For a few minutes just after dawn she sends out a pheromone lure in the hope of attracting a male.

Mantises have a well-earned reputation for sexual cannibalism, with females routinely eating suitors. So, in this case, the female is sending two messages in chemical form: 'Do come, and I promise not to eat you.'

Banded Demoiselle

Calopteryx splendens | **Odonata / Calopterygidae**

There are few things as joyous as seeing clouds of these winged sapphires catching the early-summer sun.

Winter slows our rivers, streams and canals into a lazy torpor, but with the coming of spring they slowly bubble back into life, starting under the surface. A banded demoiselle larva ascends to the surface and crawls out of the water, up a reed or grass stem. Once settled, it begins the process of drying out and hardening its outer shell, for inside it is now a fully formed adult that is squished far too tightly in and needs to get out asap. It does this slowly, splitting the casing and then easing itself out; once its wings are fully expanded, it makes its first diaphanous ascent into the sky. The males emerge first, and they are unmistakeable with their bright, metallic-blue bodies and forewings bearing a broad, black band. They maraud the water's surface and surrounding vegetation, awaiting the emergence of a female, gathering in large numbers that create a glittering spectacle.

Saint Mark's Fly

Bibio marci **| Diptera / Bibionidae**

The St Mark's Fly is named after the season in which it swarms.

There are not many insects around the world that are associated with a particular day in the year, but this hairy black dancing fly is named after St Mark's Day, 25th April, when it is said to emerge – although it is often earlier, in March. The British nature writer Roger Deakin described the nuptial dance beautifully. 'Their flight is jerky and uncertain. They kept taking off [as if] on a maiden flight, dropping out the sky quite suddenly, only to catch themselves, as if on an invisible safety net, and set some new and equally aimless course.'

The Flies With Inflatable Heads

Diptera / Schizophora

Some flies inflate their head capsule to help lever themselves out of their pupal case – pretty cool, eh?

Flies undertake complete metamorphosis, incorporating a pupal stage in which the larva forms a cocoon, called a puparium, around itself and more-or-less biologically rearranges itself over a period of time, emerging with a very different appearance from its adult stage. Emerging from the puparium is tricky – the new adult is soft and vulnerable to damage as it pushes through the hard pupal casing. However, some flies have evolved a way to ease this process: the Schizophora are a large group of flies (which include the familiar houseflies and blowflies) that have a ptilinum – a large inflatable sac on their heads. When an adult is ready to emerge, it inflates the ptilinum by pumping it full of haemolymph (insect 'blood'). There is a seam in the pupal casing which, when this pressurized water balloon presses against it, splits open, allowing the fly to exit. Once free, the fly still has a massively engorged head, but the ptilinum gradually deflates and disappears as the haemolymph is redistributed back into the fly's body.

BLACK ANT

Lasius neglectus **| Hymenoptera / Formicidae**

Some ants pass on immunity to infection to colony members.

Ants, which live in large colonies, are prone to outbreaks of disease, which in such confined spaces and with high levels of contact, can spread very quickly. The ant *lasius neglectus* partly gets round this problem by programmes of inoculation!

It isn't standing in line to receive a jab, but studies have shown that individuals that are infected by a potentially deadly fungus will offer a minute sample of it to their nest-mates, which are all siblings, mandible to mandible. The latter acquire a low-level infection, which doesn't kill them, but builds up their immunity for the future.

Bombardier Beetles

Coleoptera / Carabidae

If you think farts can be bad, you don't want to be in the firing line of this beetle...

Many insects come with their own inbuilt chemical weaponry that can be deployed in an instant to disarm predators. It could be a sting, a bite, or an excretion. Bombardier beetles have adopted a rather dramatic strategy, which is to shoot hot acid out of their bottoms. Within the tip of the abdomen of a bombardier beetle there are two separate chambers, each containing a chemical compound: hydroquinone in one and hydrogen peroxide in the other. When threatened, the beetle funnels the compounds into one reservoir, creating a volatile reaction; the mixture heats up to almost 100°C (212°F) and also produces a gas that ejects the liquid out of the beetle with explosive force. Anything in the blast radius is sprayed with near-boiling, corrosive liquid, painful to humans and lethal to predators of a similar size.

Buff-tailed Bumblebee

Bombus terrestris **| Hymenoptera / Apidae**

Bees just wanna have fun.

Almost everybody loves these big, furry, buzzy pollinators. And now there's a reason to love them even more. A study published in 2022 shows that buff-tailed bumblebees play.

What they seem to love is wooden balls that they can roll. Several experiments found that individuals rolled them even when they didn't need to, and there was absolutely nothing to be gained from it, except, presumably, for 'feeling good'.

Still more lovably, young bees played more than grown-up bees. Wonderful.

Wombat Flies

***Borboroides* spp. | Diptera / Heliomyzidae**

Wombat poo just might be the world's most extravagant niche habitat.

However terrible the world might seem at times, at least there are still wombats, wombat flies, and entomologists who study faeces to find flies. These little-known dipterans are found only on wombat dung and, of course, since wombats are unique in producing cubical dung, the flies are the only insects that are drawn to poo of this shape. Isn't that wonderful? An Australian entomologist called Dr David McAlpine is the hero of the wombat fly story. He borrowed wombat dung from a zoo, put it down near a wombat burrow and, after no time at all, found two new species for science!

Bumblebee Plumehorn

***Volucella bombylans* | Diptera / Syrphidae**

Don't be fooled by its fluffiness – this is a bee-mimicking hoverfly.

If imitation is indeed the highest form of flattery, then bumblebees should be feeling incredibly smug, for their mimics are numerous, especially among the flies. Bumblebees have a sting – a weapon that flies have not acquired. In lieu of this, flies have come up with another cunning plan; the acquisition of physical characteristics that make them look like bees. Batesian mimicry describes an animal with no natural defences that evolves the appearance of a more dangerous one in order to deter predators. A fine example of this is the bumblebee plumehorn, which comes in several colour forms that closely match different species of bumblebee. It is a large, fluffy hoverfly that uses its disguise not just to avoid being eaten, but also to sneak into the nests of the social bumblebees it impersonates. Hiding in plain sight, the bumblebee plumehorn lays its eggs in bumblebee nests, where its own larvae unfussily scavenge on detritus, and are even partial to the bee larvae developing within.

Shrill Thorntree Cicada

Brevisana brevis **| Hemiptera / Cicadidae**

Few would be surprised to hear that cicadas are among the world's loudest insects.

Have you ever woken up in the night wondering which is the world's loudest cicada? If so, help is at hand. It's a close-run thing, but the winner is the shrill thorntree cicada of South Africa, which should be called the *smart* thorntree cicada, because it is most attractive, with a smart W-shaped mark on the thorax and white spots on the wings. The male's song has been measured at 106.7 decibels at 50cm (20in) distance, just ahead of a North American and Australian competitor. This is about as loud as a chainsaw, and only slightly more musical.

Striped Horsefly

***Tabanus lineola* | Diptera / Tabanidae**

Horseflies are justifiably famous for their remarkable eyes.

Horseflies don't have a great reputation. They are biting flies that, unlike some species, approach silently. The bite is also painful. There is a Scottish expression that goes like this:

Whaur the midgies mazy dance,
Clegs dart oot the fiery lance.

The flies make a puncture in the skin of a human, or cow or horse, inject anticoagulant and then drink the pool of blood that is formed. It is only the females that do this, and they do so to acquire enough protein to form their eggs.

But against all the odds, we should like these insects. Have you seen their *eyes*?

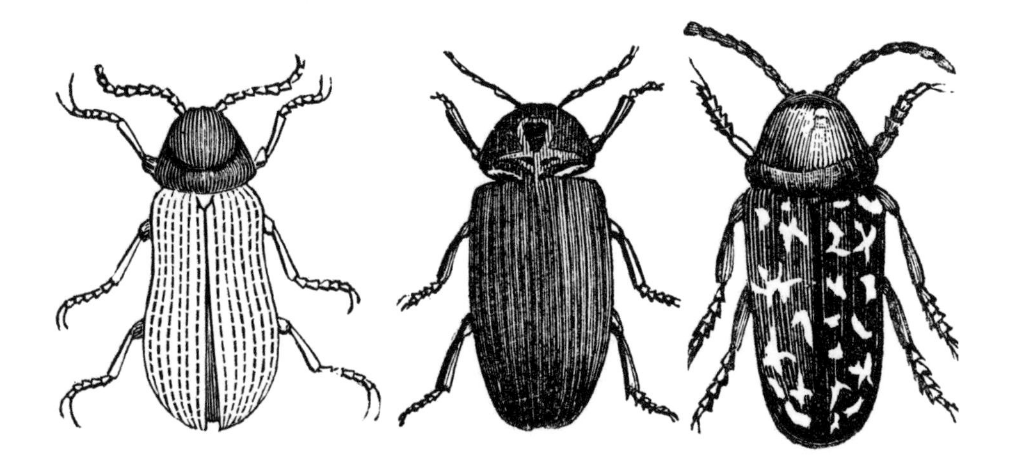

DEATH WATCH BEETLE

Xestobium rufovillosum | **Coleoptera / Ptinidae**

Tap, tap, tap... Who's looking for love in your rafters?

This tiny beetle, though rarely seen, has firmly embedded itself into the fabric of British and American folklore. The death watch beetle lives within dead timber, which from medieval to Georgian times, made up the fabric of most houses, and became a 'home within a home' for a multitude of tiny lodgers, where they burrowed tiny little corridors within the timber. It is difficult to find a mate when you are confined to the internal structure of wood, so the beetle, romantic little soul that it is, vigorously headbutts the sides of his wooden corridors to alert the potential mates to their presence. Remarkably, this cranial action is audible to the human ear, and never more so than in the dead of night, or when the house is holding a silent vigil for the dying; an event that was all too frequent in times of old. What was superstitiously transformed into the ticking clock of imminent death was, in fact, nothing more than the amorous head bashing of a tiny, sex-obsessed beetle.

BLUE-WINGED HELICOPTER

Megaloprepus caerulatus | **Odonata / Coenagrionidae**

This huge damselfly is found in the canopy of tropical forests

Damselflies are often thought of as the more delicate and vulnerable of the two groups of Odonata, their relatives the dragonflies having far greater flight speed and power. However, there are a group of damselflies in the forests of Central and South America known as the giant damselflies, and they have the largest wingspan of either of the two groups. One is the magnificent blue-winged helicopter, which lives in wet forests and lays eggs in pools of water that collect on trees. As befits a giant damselfly, it has an ambitious diet. It flies to the webs of spiders and plucks the arachnids from their silk traps, grabbing them in its forelegs and then removing all eight limbs before consuming them.

THE CATERPILLAR

Life is one big picnic when you're a caterpillar.

Insects that pupate, such as moths and butterflies, seemingly live two entirely different lives; one as a slow, tubular eating machine and the second as an ephemeral, flying progenitor. Both of these 'lives' are fraught with danger, a detail not overlooked by Christina Rossetti, who marvelled in the details of nature and its processes:

Brown and furry Caterpillar in a hurry;
Take your walk to the shady leaf or stalk
May no toad spy you, May the little birds pass by you;
Spin and die, To live again a butterfly.

CHRISTINA ROSSETTI (1830–1894)

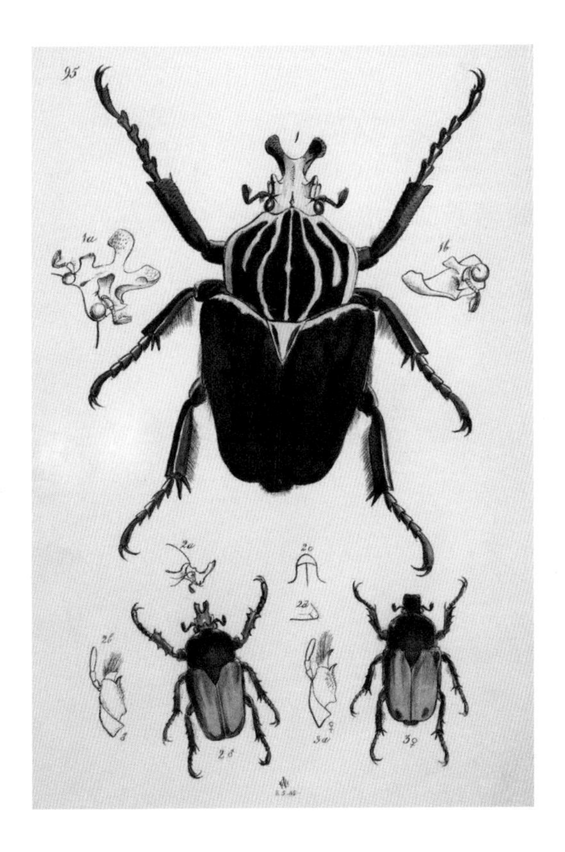

Goliath Beetles

Goliathus spp. | Coleoptera / Scarabaeidae

The insect equivalent of a blue whale, Goliath beetles are truly colossal among their kind.

Insects have evolved over millions of years to be small, and this is an adaptation that seems to work incredibly well for them. There is a limit to how big an insect can be because of the way their bodies perform gas exchange and respiration functions; being small and lightweight also helps with flight capability and mobility. The Goliath beetle, as the largest, heaviest beetle currently described, didn't get this memo. Unusually, males are larger than females, reaching up to 110cm in length. It is also one of the heaviest insects too so getting into the air does takes effort. The beetle flicks open its elytra and deploys its wings, which thrum loudly as it takes off. It bobbles about clumsily in the air as it fights to gain momentum and direction before moving off like a helicopter. Once properly airborne, the Goliath beetle is capable of flying some distance – a superb sight to behold and a tantalizing glimpse into an ancient time when many of our insects were much larger.

Giant Bull Ant / Bulldog Ant

Myrmecia gulosa **| Hymenoptera / Formicidae**

Bulldog ants are as grumpy as they are large, so don't wind them up!

The giant bull ant, sometimes referred to colloquially as the 'hoppy joe', is one of the largest ant species in the world. Adult workers can reach 4cm (1½in) with large, extended mandibles making up a good proportion of this impressive length. It is one of the bulldog ants, a group that is present across Australia (though this species is restricted to the Eastern regions). Known as 'primitive' ants, they are particularly aggressive towards prey and even each other, and some workers lead unusually solitary lives. They have very large eyes compared to many other ants so their vision is very keen when hunting and scavenging, and those huge mandibles are deeply serrated to grip and crush prey, such as bees and even other ant species. The other end of this behemoth can deliver quite a shock too – their sting is shockingly painful and irritating and can last a couple of days. Given its grumpy predisposition, this ant is probably best given a wide berth.

GREEN OAK TORTRIX

Tortrix viridana | Lepidoptera / Tortricidae

Green is a rare colour among moths, but the green oak tortrix is superabundant in European oak woods.

The green oak tortrix is an abundant moth of oak woods throughout Europe. Its caterpillars can be so overpowering in number that they can completely defoliate a mature tree. The larvae are able to spin their own threads and 'balloon' from one tree to another, and on still days in May it is easy to walk through a gossamer curtain of silk made by thousands of vacationing caterpillars. This moth is incredibly important to one of Europe's commonest and loveliest birds, the blue tit (*Cyanistes caeruleus*). The birds time their breeding to coincide with the glut of this and winter moth (*Operophtera brumata*) larvae, which they feed to their chicks at a rate of up to 1000 a day. As the oak bud bursts and the larvae begin to feed and grow, the bonanza begins.

Butterfly Flutterer

***Rhyothemis fuliginosa* | Odonata / Libellulidae**

The astonishing iridescence of the butterfly flutterer's wings is a key component of courtship.

One of the most stunning dragonflies of the South Asian wetlands must surely be the butterfly flutterer. This is an English name given to a dragonfly with several common names within its native range. It is a stocky species of the chaser family, with an inky blue body and dark, pearlescent eyes. Its wings are strongly iridescent; layers of structural colour scatter the light across them creating a hypnotic spectrum of colour, from pale pink, through purple to deepest blue. Males and females have slightly differing wing patterns, to allow them to distinguish each other as they fly effervescently through their breeding grounds.

Dance Fly

Empis opaca **| Diptera / Empididae**

Dance flies are predators, but often take time off for a spot of nectaring.

Dance flies are highly predatory on other flying insects. Their survival depends on catching awkward food, so perhaps it is not surprising that, when it comes to mating, a male can show off his wares by presenting a nuptial gift of – you've guessed it, a piece of tasty, freshly killed invertebrate flesh. Males carry their spoils into a swarm of their own species. Once there, they perform an impressive, pendulum-like, side-to-side dance. They are routinely met by a posse of excited females, and very soon the gift is delivered, with the payoff of copulation. If you think this might all be too easy, you are right. Males sometimes dupe the females by collecting dud items such as the fluffy seeds of dandelions or willows and passing these off as edible gifts. Many a female avoids such duplicity, but a remarkable number are fooled. They copulate with the feckless males, despite getting nothing in return.

May Bug (Cockchafer)

Melolontha melolontha | **Coleoptera / Scarabaeidae**

The brush-like antennae of the May bug are worthy of an expensive fashion accessory.

You have to root for this marvellous character; it belongs in a gentler world, one without artificial lights to disorientate a barely airworthy, bumbling beetle. It flies with a hum and bad steering, and may clatter into almost anything, whereupon it often falls to the ground on its back, legs flailing. Close up it sports delightful, brush-like antennae, with seven 'leaves' on a male's and six on a female's.

Cockchafers have a penchant for many types of vegetation, including root crops, and in the past appeared in such numbers as to cause some economic problems. Hence in a court at Avignon, France, in 1320, local May bugs were commanded to retire to a designated area on pain of death, an example of the bizarre medieval practice of animal trials.

Peacock Butterfly

Aglais io | **Lepidoptera / Nymphalidae**

The peacock butterfly has an impressive array of defence strategies, including bird mimicry and hissing.

There's something rather iconic about the peacock butterfly where I come from; it is one of UK's most eye-catching species. Those panels of scarlet, punctuated by interrogative blue eyes cut through the late spring gloom, and announce winter well and truly over. The peacock's startling visual appearance may please us greatly, but it has an altogether more practical function. The false eyes on each wing are an adaptation to fool predators into thinking that the butterfly is much bigger and scarier than it really is; in fact, if you rotate a peacock 180° it looks like an owl – its body resembling a beak between hungry, piercing blue eyes on the hindwings. And if that isn't enough to deter predators, if threatened further it will rapidly rub its wings together to produce a hissing noise that is audible to our ears and will make larger assailants think twice.

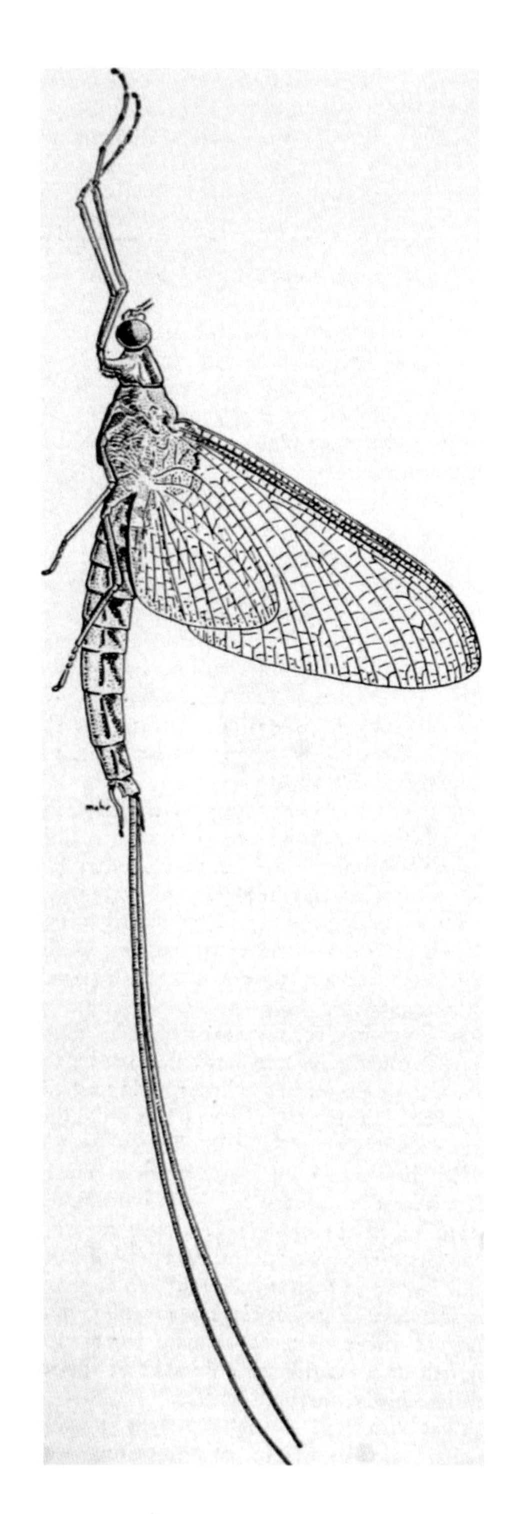

American Sand-burrowing Mayfly

***Dolania americana* |**
Ephemeroptera / Behningiidae

The one thing that most people known about mayflies is that 'they only live for a day'. It isn't strictly true, of course, because in their immature stages they may persist for a few years. But the reproductive stage is often very brief and measured in hours.

In the case of the American sand-burrowing mayfly, however, a single hour would be stretching it. Male nymphs emerge, moult and mate, all within about 30 minutes in the hours before dawn. The female nymphs emerge after the males. They mate and lay eggs in the water before dying, all within a barely credible 5 *minutes*. It is the shortest adult lifespan of any insect. And none of them see daylight.

Bloody-nosed Beetle

***Timarcha tenebricosa* | Coleoptera / Chrysomelidae**

Above: You can't fail to be charmed by the wonderfully charismatic bloody-nosed beetle, which is far more lovable than its name suggests.

Opposite: Individuals of this mayfly may only live in the adult stage for 5 minutes.

Despite its macabre name, the bloody-nosed beetle is one of the most charming insects you will ever see. This rotund, black beetle may not be colourful or shiny, but what it lacks in glamour, it makes up for with personality. In late spring and summer, it can be spotted across the Palearctic, on bedstraws or trundling in earnest across paths and meadows. The male's stout little legs have large, heart-shaped tarsi ('feet') with short, orange pile on the underside; these are used to grip onto the back of the significantly larger female when mating. Its rather odd name derives from a super-smart defence mechanism; when threatened it can discharge its own haemolymph – a red liquid – from membranes in its mouth. Not only is the liquid distasteful, it also gives the impression of having an explosive nosebleed (except for the small detail that beetles don't have noses), startling predators into retreat, two-fold.

Rose Chafer

***Cetonia aurata* | Coleoptera / Scarabaeidae**

These flying emeralds are an impressive sight on early summer flowers.

Rose chafers are spectacular, blingy beetles of the Scarabaeidae family. They time their emergence with the blooming of the preferred nectar sources, which include honeysuckle, dogwood, roses and elder. They are vivid emerald green, but can sometimes look bronze due to their iridescence, also known as 'structural colour'. Alternating layers of chitin in the chafer's exoskeleton reflect daylight into our eyes, and our brains perceive the metallic green colour. However, under harsher lighting conditions such as camera flash, the structural layers bounce light a different way, eliminating the reflected green and making the rose chafer appear deep metallic bronze – a phenomenon common in many beetles. Nesting habits of the rose chafer are fairly cosmopolitan; compost heaps, deadwood and loose, humus-rich soil are all attractive egg-laying opportunities for females, and the larvae are highly effective recyclers of dead organic matter.

Emperor Moth

Saturnia pavonia | Lepidoptera / Saturniidae

The emperor moth has an impressive array of wardrobe changes throughout its life.

Many moths have at least one drab phase either as larva or adult. The emperor moth, however, is a spectacle at all stages of its life. The larva is tiny, black and covered in spines. As it grows, it develops orange spots to accompany the funky spines. The spots gradually radiate and join up into broad bands as the caterpillar's girth increases. The final moult sees another radical wardrobe change, as the increasingly orange caterpillar becomes neon green. Presumably these sudden changes serve to continually confound predators, leaving them in doubt as to whether their prey is safe to eat – staying one step ahead of the opposition, if you like. But the emperor moth isn't finished there, because it doesn't leave its flamboyance behind in the chrysalis – it brings the dressing-up box with it and emerges with stunning patterns and false eyes; the female is more monochrome whereas the male opts for a colour palette that would work beautifully in a 1970s hotel lobby.

Jet Ant

Lasius fuliginosus | **Hymenoptera / Formicidae**

Jet ants have taken nest-building efficiency to the next level, by covertly taking over the nest of another colony.

Rather than going to the trouble of establishing a colony of its own, the jet ant hijacks an existing colony. This behaviour is surprisingly common in the ant world. When a new queen emerges: instead of scouting out her own little territory to settle in, she sets off in search of occupied real estate, usually belonging to another species of black ant. Now for the tricky part – you can't just walk into an ant nest; the fiercely territorial workers will swiftly dispatch any interlopers. The solution for this is to disguise oneself, so the trespassing queen kills a worker or two and acquires their scent. Suitably masked, she can now walk through the nest, right into the invested queen's chambers. With brutal sleight-of-hand she kills the queen, assumes the throne, and lets the workers continue their duties, none the wiser to the fact that they serve a new monarch. The usurper begins to churn out her own eggs and before long the original colony is completely replaced by a population of shiny new jet ants.

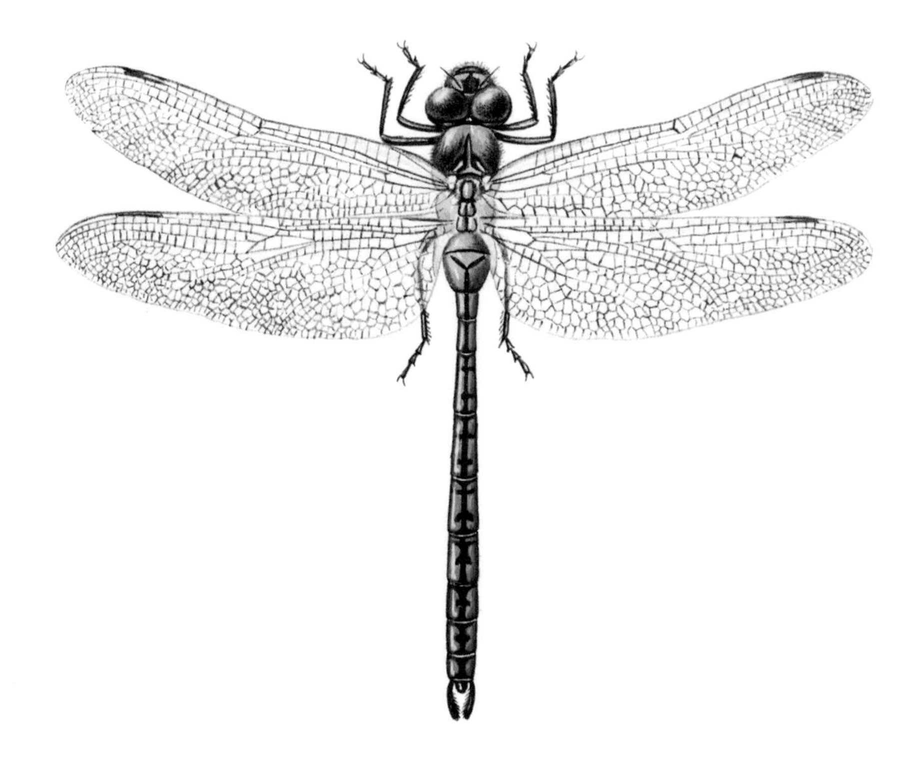

EMPEROR DRAGONFLY

Anax imperator **| Odonata / Aeshnidae**

Beautiful but deadly, the emperor dragonfly homes in on prey with astonishing accuracy.

The emperor is one of those big, confident, fast and impressive dragonflies that come out in the summer. It spends much of its day patrolling back and forth along the edges of well-vegetated waterways, sometimes rising high into the air with bursts of speed to catch flying insect prey. It is raw power and energy within a glittering, shimmering livery of bright colours.

Studies have shown that large dragonflies are the most efficient predators in the animal kingdom. The apex predators that we know and love such as eagles, lions, wolves or orcas all are outperformed by these everyday insects, which have an incredible 95 per cent capture rate.

THREE-BARRED SOLDIER

Oxycera trilineata **| Diptera / Stratiomyidae**

This chartreuse green soldier fly contrasts fabulously against the more muted palette of its marshy habitat.

It's a warm, sunny day in midsummer, and the sun is radiating across floodplain grassland. The air hums with the sound of busy insects, all going about their lives; you watch them zip around, occasionally landing for a brief moment. Most of them will be flies and bees, dark grey or brown in colour and fairly uniform, though some of them will display the odd flash of yellow, or have snazzy markings. One little fly, however, didn't get the memo about dressing conservatively; it doesn't just make an entrance – it knocks the door down. The three-lined soldier is the most bonkers shade of green it's possible for an insect to be. It's hard to actually describe the colour (lime/viridian/chartreuse?), as it is so vibrant. And it's not just the body; the eyes are also a dazzling green, but because they are iridescent, they appear to change from neon green to coppery red depending on how the light shines on them. The striking horizontal band through the centre of each compound eye makes it all the more mesmerizing.

YELLOW DUNG-FLY

Scathophaga stercoraria **| Diptera / Scathophagidae**

The attractive yellow dung-fly perched upon its insalubrious dancefloor.

Does 'scramble competition on a cowpat' sound appealing to you? Welcome to the world of the yellow dung-fly. The advance plan is that girl meets boy on fresh dung, where she will lay eggs. The reality is that she meets lots of boys, all at once, and there is unseemly competition between them. As they scramble for access, she can be collateral damage, sometimes being injured and at best being pressed down into the excrement. If all goes well, she will copulate with her chosen male – and get out alive.

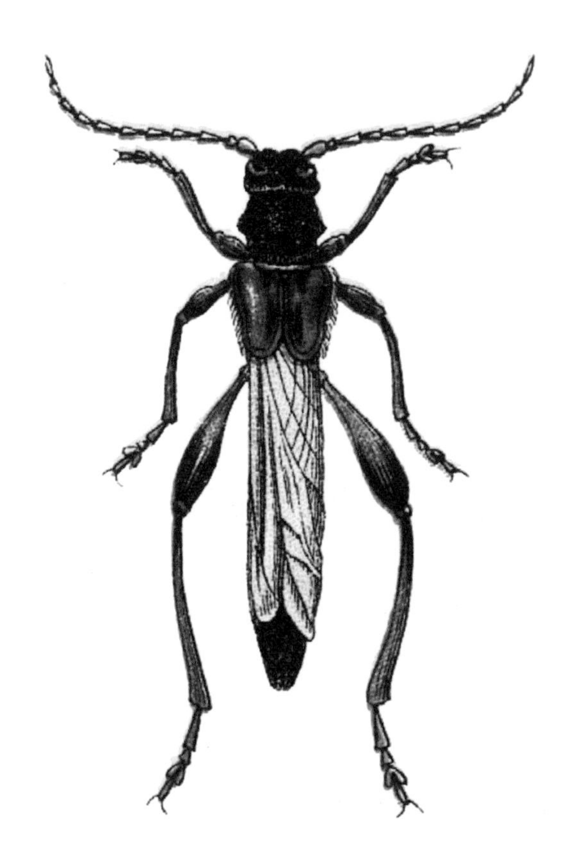

Wasp-mimic Longhorn

***Necydalis major* | Coleoptera / Cerambycidae**

The shortened wing cases of this longhorn beetle expose the wings, only further enhancing its mimicry skills.

The forests of eastern Europe and north Asia are home to some weird and wonderful beetles. *Necydalis major* is an unusually slim longhorn beetle with a constricted waist which, with its two-tone colouring and shorter antennae, gives it more than a passing resemblance to a parasitic wasp. The disguise is further enhanced by the wings – usually covered by the elytra in beetles, but in this case the elytra are greatly reduced, leaving most of the hindwings exposed, just like a wasp. Like most longhorn beetles, the larvae of *Necydalis major* live in, and eat, the wood of deciduous trees, taking 1–3 years to reach pupation; this species appears to favour alder, lime, oak and poplar.

Bulgarian Emerald

Corduliochlora borisi | Odonata / Corduliidae

Remarkably, the Bulgarian emerald was only discovered in 1999.

Many people would be astonished to know that a new species of dragonfly could be discovered in Europe in recent times. However, it was in 2001 that Dr Milen Marinov managed just that, coming across a beautiful and distinctive dragonfly patrolling along a river in the Rhodope Mountains in Bulgaria. It is still only found along the shady parts of a handful of slow-flowing rivers and is already threatened. There are estimated to be only 9,000 individuals in the world.

Common Scorpionfly

Panorpa communis **| Mecoptera / Panorpidae**

Not quite a fly, but not a scorpion either. Scorpionflies are part of a wonderfully unique group of insects that possess attributes of other insect orders.

Late spring hedgerows and woodland edges are coming to life with the sight of butterflies, bees, flies and other familiar insects. Then you spot something sitting on a leaf that doesn't seem to fall into any category; a glorious mash-up of other stuff that must mystify even its six-legged neighbours. *Panorpa communis* is resident to Europe and northern Asia. It has two pairs of wings, like bees and wasps, which it holds at rest in a similar way too, but its slender, delicate outline is distinctly fly-like. The long face looks more like a beak. Down the other end, though, is the real surprise, for males have genitalia that look just like a scorpion's tail. Fear not, for the scorpionfly cannot sting, it is for display and practical purposes only. Females look very similar in all other ways, but lack this curved, bulbous extremity. The larvae are also rather clever – they look like predatory beetle larvae, which gives them some protection when sifting for their own food through the leaf litter.

APHIDS

Hemiptera / Aphididae

The reproductive possibilities of aphids are barely imaginable – without predation they would easily fill the Earth in a few weeks.

Aphids are the familiar very small bugs that suck plant juices; they are also called greenfly. They aren't particularly popular insects, but they are very successful, truly remarkable and their reproductive capacity is frightening. It has been estimated that if, at the start of spring, the offspring of a single female all survived, she would have a quadrillion by the end of the breeding season. A large field of crops may host four metric tonnes of aphids.

The way they do this is almost eerie. At the beginning of the season, a surviving female (a fundatrix) will reproduce without the need of a male. Or, indeed, of eggs; she gives birth to live young. Aphids are like Russian dolls. When a female gives birth to a nymph, the latter is already pregnant with a nymph of its own!

Periodical Cicada

Magicicada spp.

Hemiptera / Cicadidae

Few insects, or any animals, have a life cycle as strange as the periodical cicadas. These animals emerge in a certain year, mate, lay eggs and then only appear again 13 years or 17 years later, depending on the species. When they emerge, they do so locally in their millions; the strategy is that there are so many at once, squeezed into a period of a few weeks, that predators cannot possibly wipe them out. They are everywhere, and very noisy collectively, reaching up to 100 decibels. And then, where they are superabundant, they die off and don't appear again for a very long time.

These days, scientists know exactly when and where certain 'broods' will hatch. They can warn people to avoid having a wedding or other open-air event during a large emergence. North Americans have been known to take cicada holidays!

Nobody is quite sure why periodical cicadas have such long growing times, and nobody knows how they manage to coordinate so well. Imagine if an individual cicada got it wrong; it would need to wait another 13 or 17 years to meet any of its kind!

Horn Fly

Haematobia irritans | Diptera / Muscidae

Above: Horn Flies can lay their eggs in what seems almost indecent haste.

Left: You need to look your best after 17 years hidden away.

The horn fly is a blood-sucking species that lives on cattle. Adults may live their entire adult lives on one individual, having first flown up to 16km (10 miles) to find the perfect match.

The only time a female fly leaves the side and hide of its host is to lay eggs, which it does into fresh dung. And however revolting that might sound to you, you have to admit one thing. This fly is on the ball.

It takes less than 15 seconds after deposition for *Haematobia* to lay its eggs. In fact, the deed may be done before the cow has even finished its 'session'.

RED WOOD ANT

Formica rufa | **Hymenoptera / Formicidae**

The formic acid produced by wood ants has many useful applications in the human world.

The red wood ant (also known as the southern wood ant) is commonly found in the open mixed woodlands of northern Europe, where its large, domed nests, dressed in pine needles and other fallen vegetation, can be seen as bold convexes rising from the forest floor. A mini-metropolis such as this can contain hundreds of thousands of individual adults and larvae, as well as the queen (of which there can be more than one). Nests are usually close to glades or margins, where the heat of the sun can reach them; the ants will form penguin-like 'huddles' in the morning rays, to warm themselves before heading out on foraging duty. Red wood ants naturally produce formic acid, which they can expel from their abdomens in a high-velocity jet to deter predators. The acid is tangy, and tastes of lemons; it has become widely used in limescale removers, the leather tanning process and as a food preservative. It has even been distilled into the world's first insect-flavoured gin (and, yes, I have tried it...).

SOUTHWESTERN MOLE CRICKET

Gryllotalpa vineae **| Orthoptera / Gryllotalpidae**

Mole crickets live in burrows in the soil and are rarely seen above ground.

Mole crickets live up to their name, digging burrows for themselves using their greatly enlarged front legs, which really do recall those of the mammal. They also eat invertebrates that live in the soil, as do moles themselves.

Mole crickets use sound to advertise their presence to a mate. They stridulate underground by rubbing the rear edge of the left forewing against ratchet-like teeth on the lower edge of the right forewing. However, their burrow is Y-shaped, with two entrances/exits that act like musical horns. The structure of the burrow and the paired openings greatly amplify the signal, which is one of the loudest among all insects, and can be heard by the human ear at least 200m (650ft) away.

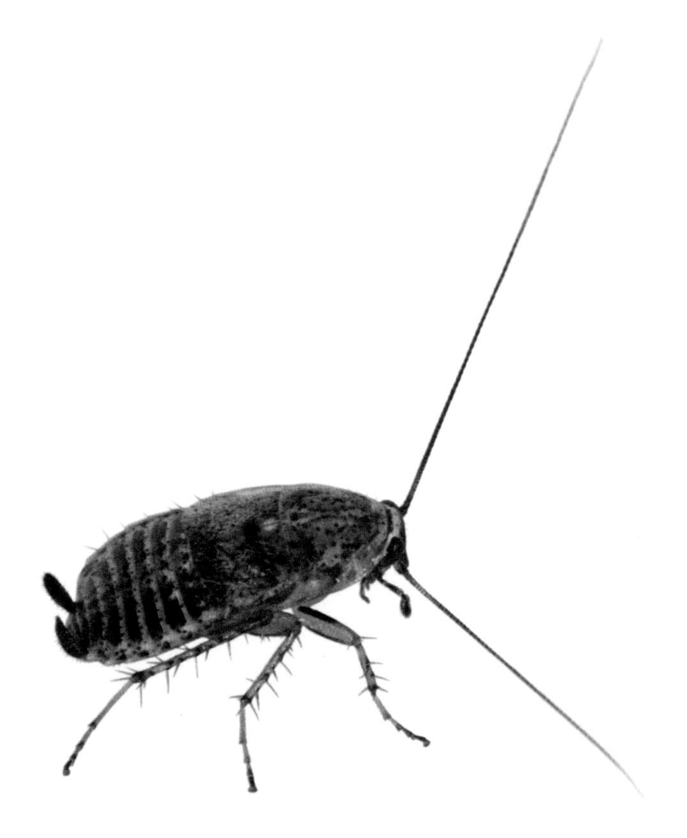

Lesser Cockroach

***Capraiellus panzeri* | Hemiptera / Ectobiidae**

The lesser cockroach is a small, reclusive animal that prefers to stay far away from us.

The mere mention of the word 'cockroach' is enough to strike fear into the hearts of many, for it is a word synonymous with domestic infestation and chemical warfare. But let us for moment cast such aspersions from our minds and turn our gaze instead to the lesser cockroach, a substantially smaller, decidedly human-averse, and wholly more lovable creature. The lesser cockroach is one of several species of cockroach that are native to the UK, but which most of us will live our lives without ever seeing. It is an exquisite little bug, with finely etched wings and large, enquiring eyes. The males are fully winged, whereas the females are short-winged, revealing a dark brown abdomen with pale markings – it might be mistaken for a woodlouse at first glance. It is usually found in sandy, coastal habitats, where it scuttles around, scavenging on a wide range of plant and animal matter.

Pride of Kent Rove Beetle

Emus hirtus | Coleoptera / Staphylinidae

This fabulously fuzzy rove beetle does a great job of looking like a bumblebee.

If you are fortunate enough to see this fabulous beast fly past you, you would be forgiven for thinking it was a bumblebee, with its bright yellow stripes; even walking around on the floor it has a distinctly bumbley vibe to it. This highly distinctive fur-ball is one of the largest rove beetles, a huge family of beetles that are known for being highly efficient predators. It is found throughout north and west Eurasia, a frequenter of livestock pastures, where it hunts for small invertebrates among horse and cow dung, dismembering them before consuming them.

Common Froghopper (UK)
Meadow Spittlebug (USA)

Philaenus spumarius **| Hemiptera / Aphrophoridae**

The young of this bug live in foamy protective secretions that look like saliva.

This is a bug that packs a punch. It is most famous for its ability to produce foamy domiciles to protect its young. The homes look like little splodges of spittle, hence the name meadow spittlebug. They are perfect protection for the eggs and nymph because they don't look edible and are, indeed, slightly acrid. Famously, the splodges are made by the froghopper inserting its rostrum into the xylem (the tissues that transport water and minerals) of a plant and expelling the sap through its rear end under pressure; two 'blow holes' on the abdomen add the air to create the foam.

This remarkable insect holds two records: it can jump 100 times its own height; and it has been known to feed from no less than 1,300 species of plants.

Coastal Silver-stiletto

Acrosathe annulata | **Diptera / Therevidae**

A fluffy flash of silver in the sand signifies the presence of this handsome fly.

One of the most surprising things about insects is how hairy they can be. Hair is a loose term in itself, as insects do not possess keratin-based hair or fur, as mammals do; instead their 'hair' is modified chitin called setae, which grows from the cuticle. Setae are connected directly to the central nervous system and relay valuable environmental data from an insect's surroundings to its brain. Some setae are sparce and bristly; some are short, dense and fuzzy; and others, such as the coastal silver-stiletto, are all-in for the polar bear look. It is just the males of this species that have evolved whole-body fluff; females are relatively hair-free. You'll usually have to go to the seaside to find them, as they are strongly associated with sandy habitats. Here, they will sit on open sand in the sun, their furry pile gleaming against the ochre sand.

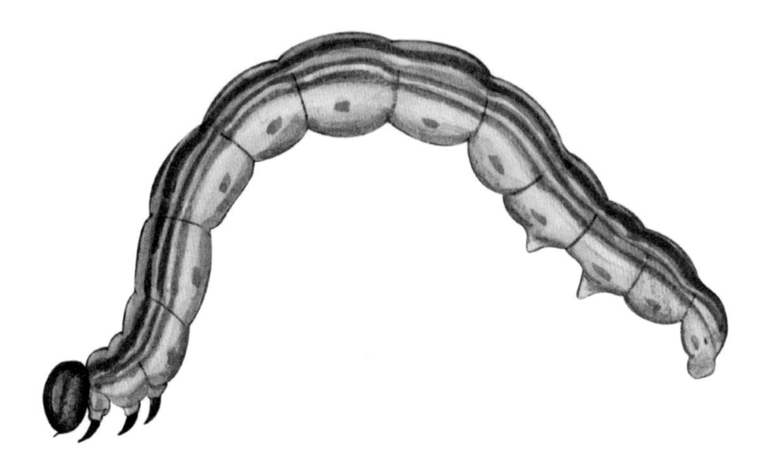

INCHWORMS
Lepidoptera / Geometridae

The caterpillar of a Gemoetrid moth 'measuring' its progress forwards.

Have you ever wondered what an inchworm is? It does exist. The name refers to the caterpillars of moths in the huge family Geometrid. The name means 'earth-measurer' and other names include measuring bugs and loopers. By now you may have guessed that these caterpillars move in a distinctive way. They have true legs at the front, three pairs of prolegs at the back (instead of five, as in most caterpillars) and nothing in the middle. That means, to move forwards, they must hold fast with the back legs while inching along, then hold fast with the front legs as they bring their back legs up to catch up, looping the body.

Do they have time to appreciate the plants they eat? As the song 'Inchworm' suggests (written by Frank Loesser and featured in the 1952 movie *Hans Christian Andersen*), maybe not.

Inchworm, inchworm (two and two are four)
Measuring the marigold (four and four are eight)
You and your arithmetic (eight and eight are 16)
You'll probably go far (16 and 16 are 32)

Inchworm, inchworm (two and two are four)
Measuring the marigold (four and four are eight)
Seems to me you'd stop and see (eight and eight are 16)
How beautiful they are (16 and 16 are 32)

Alkali Fly

Ephydra hians **| Diptera / Ephydridae**

Adult alkali flies have a highly unusual ability to submerge, and even feed underwater.

Very few adult flies can survive underwater, let alone thrive in it. The alkali fly resides in the saltwater lakes of North America. If the larva is not unusual in its aquatic lifestyle, the adult certainly is; it can submerge to the bottom of the lake and stay there for some time without drowning. How? By making its own aqualung, for this fly is superhydrophobic; it has an unusually hairy body that repels water, and this is further enhanced by a waxy coating that the alkali fly applies to itself. When it enters the water the hairs and wax trap a layer of air around the fly, encasing it in a bubble, within which it can breathe freely. But there is further ingenuity – the fly has large claws that grip to substrate without bursting the bubble, and the bubble seals itself around the outer rim of the eyes, leaving the vision unimpeded. The fly can then potter happily around underwater in its scuba suit feeding on algae and laying eggs. When it leaves the water the bubble bursts, leaving this incredible fly perfectly dry and ready to fly.

WESTERN BANDED GLOW-WORM

***Zarhipis integripennis* | Coleoptera / Phengodidae**

By day the Pacific Northwest hosts stunning landscapes, by night the western banded glow-worm puts on a captivating display.

On calm nights in the Pacific Northwest, male western banded glow-worms are attracted by the light given off by females. And no wonder; the females resemble the larvae, being sausage-shaped, and rings of light glow from between the segments – a most singular, intriguing sight. The males attend in numbers.

In this species, males greatly outnumber females. To a male, finding a partner with which to mate is all but a once-in-a-lifetime opportunity. So, the glow-worm dancefloor becomes a bloodbath. The males fight with their vicious jaws and the body parts mount up. '*Et tu, Brute*?' You bet.

HUMMINGBIRD HAWK-MOTH

Macroglossum stellatarum **| Lepidoptera / Sphingidae**

Reports of hummingbirds in Europe are invariably the eponymous hawk-moth.

This moth truly is a hummingbird tribute act. Convergent evolution has produced an insect that hovers at flowerheads in the daytime and drinks nectar, using a long (up to 25mm/1in) proboscis instead of a bill. The wings even hum as they are beaten up to 85 times a second, and the hawk-moth has the same kind of dashing flight as the glittering birds – it can fly at 19km/h (almost 12mph). Many people are genuinely fooled when they see one.

The hummingbird hawk-moth is a Eurasian species that prefers warmer climates, for example around the Mediterranean, but is also becoming commoner further north as the climate heats up. It used to be much less common in the UK, for example, than it is now. It was so notable that, when servicemen spotted a group flying over the English Channel on 6th June 1944 (D-Day, when the allies began their assault on German-occupied Europe in Normandy, France), it was considered a sign of good luck.

Red Soldier Beetle

***Rhagonycha fulva* | Coleoptera / Cantharidae**

The florets of an umbellifer are the happy place for many insects.

There are certain things that entomologists notice that other people seldom do. For example, you know those white flowers with branching florets that make an umbrella-like top, commonly known as umbellifers? Take a look at one. You'll see that it's a dancefloor for insects, a place where they come to breed, to feed and to bleed. It's a microcosm of life.

In Europe, one species is particularly well known for its love for these flat-topped flowerheads – the red soldier beetle. Its activities are so blatant that it's been nicknamed the hogweed bonking beetle, so frequently is it seen in mating pairs. For its brief life, the flowerhead is its world.

KNOPPER CHALCID

Ormyrus nitidulus | Hymenoptera / Ormyridae

A female *Ormyrus* wasp investigates emerging acorn buds, detecting cynipid wasp larvae into which she will lay her eggs.

Have you ever wondered about that strange, crinkly growth on an oak tree where an acorn should be? It's a knopper, the handywork of *Andricus quercuscalicis*, a parasitic wasp that lays an egg under the outer surface of an acorn, causing a biochemical change into a wrinkled form. Is this terrible news for the oak tree? Well, in this case *Ormyrus nitidulus*, a beautiful wasp with breathtaking metallic turquoise and black stripes, brings a fascinating twist to this tale. *O. nitidulus* is a chalcid, a family of wasps which lays its eggs within the bodies of other parasitic wasps; it is a hyperparasitoid. The female can sniff out the developing gall wasp within its knopper gall, then insert its ovipositor through the gall tissue and into the *A. quercuscalicis* egg or larva with pinpoint precision. The *nitidulus* larva then eats its host and emerges from the knopper to repeat the cycle. We can see these tiny beasts flitting around oak trees from early summer, but it's hard to see, on a human scale, just how spectacular these wasps are.

GIANT STONEFLY
Perla grandis
Plecoptera / Perlidae

If you see any stonefly, it's a good sign. These primitive insects, along with caddisflies (Trichoptera) and mayflies (Ephemeroptera), make up the so-called EPT index, which is used the world over for assessing water quality. Stoneflies are entirely intolerant of pollution, so the Plecoptera part of the index is very important.

This impressive species occurs on rivers in the mountains of southern Europe. The last-stage nymph leaves the water at night and clings to a stone. It moults and leaves behind it a sinister exuvium, looking like some kind of alien has been among us.

GRIG KATYDIDS

Cyphoderris spp. **| Orthoptera / Prophalangopsidae**

Above: Mating in the Buckell's Grig is not for the faint-hearted.

Left: The presence of any stoneflies is a sign of good water quality.

Two closely related species of katydid live intriguing, interwoven lives. Their territories overlap, as do their lifestyles: they will even mate with each other. One difference between these species is their mating seasons – *C. buckelli* is earlier in the year than *C. monstrosa* but there is some overlap, which can result in interspecies shenanigans. But hybridized offspring are infertile, so what's the point in this? Well, to combat a lack of food resources, females have evolved to eat the wing tips of the males. The female finds a male to mate with so she can chew on his wings whilst he is distracted. Larger, fresher wings are preferred, and because male *C.monstrosa* emerge later in the year, they provide a super-fresh supply of wing when other food is scarce.

Lesser Albatross

Appias paulina | **Lepidoptera / Pieridae**

Lepidopteran albatrosses don't fly as well as their bird equivalents, but they are significant travellers.

You don't expect to see albatrosses in Sri Lanka until you realize that they are butterflies. Some Sri Lankan butterflies are big and showy and, in line with the tropical environment; many are stunningly coloured. Personally, though, my finest butterfly experience on the island involved the comparatively modest albatrosses, which are no more than black and white, sometimes with a tinge of green. One June day I was driving along the east coast, and suddenly it looked as though the weather had changed. Butterflies, in their thousands – perhaps millions – began drifting west to east across the road, like large snowflakes. They were completely unhurried, but equally persistent, driven by something internal. Some almost brushed the verges, while others flopped close to the treetops. For almost half an hour I drove through this gentle river of moving forms. Who knows where they were headed or what they would face on the way? It was a truly astonishing spectacle.

Aspen Leaf-rolling Weevil

Byctiscus populi **| Coleoptera / Attelabidae**

This tiny weevil can manipulate and roll up leaves many times bigger and heavier than itself.

Of the many examples of insects utilizing leaves for their own purposes, the leaf-rolling weevils deserve a special mention. The aspen leaf-roller can be found in temperate successional woodlands. These tiny beetles can roll up leaves considerably larger than themselves into neat cigars and, even more impressively, while the leaves are still attached to the tree. Starting on an outer edge, a female commences a first tight roll, then lays a few eggs down the length, before continuing to roll the leaf up tightly like a carpet, the eggs now nestled deeply in the centre. The leaves of suckering shoots are preferred to mature trees, probably as they are more tender and malleable. The larvae will only eat dead and decaying leaves, and so the female punctures the leaf stem with her egg tube to accelerate its death. The larvae eat their edible nest from the inside, before dropping to the ground to pupate. The adults are particularly handsome, with a metallic sheen that glows every colour of the rainbow in the aspen-rich woodland.

Grote's Bertholdia

Bertholdia trigona **| Lepidoptera / Erebidae**

Some moths can rest easy flying at night, since they can jam the radar of bats.

On gentle midsummer nights, there is conflict in the skies. Bats are trying to catch moths, and moths are trying to stop them. Grote's bertholdia, of North America, is one such moth, a type of tiger moth that shows its teeth. Scientists have shown that it has two main defences. One is an obvious one; if it detects a bat closely approaching, it simply dives down to Earth. The second one is much more sophisticated. It emits very loud pulses that have the effect of jamming the bat's radar, impairing its echolocation. Remarkably, this moth is able to detect false alarms by listening to the bat. It only uses its jamming if the bat has initiated a genuine attack.

Slightly Musical Conehead

***Neoconocephalus exiliscanorus* | Orthoptera / Tettigoniidae**

The various species of coneheads in North America differ from each other only slightly!

Today, let's hear it for the 'slightly', inspired by the unflattering epithet of this fabulous insect. It sounds, to the ears of this writer at least, as musical as many other crickets, an endlessly repeated chirp on one note. Go into a marsh or slightly untidy field corner in the eastern USA and it will add a dash of atmosphere to a night in midsummer. Neighbouring males synchronize their calling.

This cricket does bear, it must be said, a slightly longer 'nose' than its relatives, 4–6mm (around ¼in) in all. The male offers a slightly more modest spermatophore (a specialized edible gift) to the female in courtship than many other crickets do, too.

In August 2007, this species was recorded in Canada for the first time. This encounter therefore increased the known biodiversity of the country – slightly.

NORTHERN SPREADWING

Lestes disjunctus **| Odonata / Lestidae**

Northern spreadwings deposit their eggs as deep in the water as they can reach, possibly to evade being found by surface predators.

The northern spreadwing is found throughout North America in stillwater ponds and wetlands. Like many damselflies it is sexually dimorphic, meaning the sexes look different. Males are ice-blue and black, whereas females are metallic-green with brown eyes. This dimorphism may help the sexes to identify each other easily; when paired up, the male is easily distinguished as he clasps the female's head to prevent other males from mating with her. He holds onto her throughout the mating process, which involves them perching on a leaf or stem on the water's surface. The female dips her abdomen into the water to lay eggs. Female northern spreadwings appear to try and oviposit as deep as possible, and have even been observed completely submerging themselves, possibly to deposit their eggs further down than predators might be expecting and ensure the best chances of survival for their offspring.

Tisza Mayfly

Palingenia longicauda **| Ephemeroptera / Palingeniidae**

For a few short days in midsummer, mayflies cover the surface of their eponymous river.

Central Europe in mid-June provides one of the few examples in the world of an insect spectacular that attracts tourists. It is the annual *tiszavirág* or Tisza blossom, a blooming of mayflies, and it occurs on the River Tisza in Hungary.

The Tisza mayfly is Europe's largest of its kind, and it serves its time as a nymph for three years before flowering as an adult for a few hours in June. During calm, settled weather, millions of adults emerge and indulge in their mating dance, usually in the evening. They literally cover the river and the sight is truly extraordinary, if a little sad, with so many floating bodies. The hatching occurs over the course of about a week.

These mayflies are very unusual for dancing so low that they touch the surface of the water. That is because the river does not host large, predatory, surface-feeding fish.

Violet Black-legged Robberfly

Dioctria atricapilla

Diptera / Asilidae

Eurasian grasslands are beginning to hum with the sound of insects in early summer, as a multitude of organisms emerge to feed and reproduce. The result of this sudden buzz of activity is a meadow-wide, all-you-can-eat buffet, and before long carnivorous customers will arrive to sample the menu. The robberflies are supreme predators in grasslands; they fly with strength and agility through the vegetation, braking suddenly to land for a rest or to gather some warming solar power before setting off again to hunt. These are aerial ambush predators, which can snatch insects out of the air, or off vegetation with brutal precision, carrying off their quarry to a perch, where they will suck out the soft, liquid innards. The violet black-legged robberfly is glossy black, small and beautifully sleek, and, as its name suggests, it has a purple sheen in certain light. It is less stocky and hairy than many of its robberfly relatives, and is therefore easy to overlook, but once you spot it and get a closer look you will see those large, bauble eyes in the most mesmeric, blueish turquoise hue.

Flame Skimmer

Libellula saturata **| Odonata (Anisoptera) / Libellulidae**

Above: Dragonflies are a prevalent motif in Indigenous American artwork.

Left: Summer grasslands are buzzing with aerial predators such as this violet black-winged robberfly.

The dragonfly has long been venerated by the Indigenous nations of North America as a symbol of transformation, good fortune and healing. For the Navajo people, dragonflies are totemic of the purity of water. The ancient Hopi people etched and daubed representations of dragonflies into the rock faces of the New Mexico desert, and the Zuni told stories of how they saved their own ancestors from starvation. Navajo sand paintings and jewellery, and pieces by the Pueblo potters of the southern states feature dragonfly emblems. The flame skimmer is native to the desert biome of North America. The sunset-orange males and terracotta females contrast strikingly with the green vegetation, and it is quite easy to imagine this glorious dragonfly zipping through the pre-colonial American landscape, watched and mythologized by so many ancestral communities.

Predatory Fireflies

Photuris spp. | Coleoptera / Lampyridae

Flame fatale – some female fireflies eat the males they attract.

Fireflies or 'lightning bugs' are found in many parts of the world, their unique 'flashes' of bioluminescence a feature of rural places in summer. It is a somewhat romantic notion that the females should send out flashes of code to potential suitors.

But firefly biology is incredibly complex and, in typical insect fashion, utterly ruthless. Different species have different codes of flashes, sent at different seasons, different times of night and different heights. This should be sufficient to ensure that the right seeking male finds the right signalling female. However, predatory fireflies of the genus *Photuris* mimic the signals of other species (such as *Photinus*) in order to lure males to their lair, whereupon they eat them. To combat this, many *Photinus* have evolved a complex two-way, light-flashing code where both male and female flash at each other to reduce the chance of being eaten.

Elephant Hawk-moth

Deilephila elpenor **| Lepidoptera / Sphingidae**

This is the sort of character that sparks fascination with moths.

The elephant hawk-moth is a great breaker of moth taboos, a gift to entomologists who run public moth events to promote interest in these insects. People can't believe they have pink elephants in their neighbourhood!

Scientists have discovered that elephant hawk-moths have remarkable eyes, and they see so well at night that they can distinguish colours, even if there is nothing but starlight, at a luminance of just 0.0001 cdm^{-2}. That isn't very much! Humans are completely colour-blind in such an environment, as are bees and other insects. But the moths see colours, and even when the environmental illumination changes, they see the same colours, something known as colour constancy. Seeing colour at night is a vanishingly rare ability. Studies on the eye show that, in contrast to humans, who have blue, green and red photoreceptors in the eye, elephant hawk-moths have ultraviolet, blue and green.

Above: Even in the harshest conditions, insects have found a way to thrive.

Opposite top: The wings of an alderfly look like a stained-glass window.

Opposite bottom: Tiger beetles are supreme and speedy hunters.

NON-BITING MIDGES OF THE ARCTIC

***Smittia velutina* | Diptera / Chironimidae**

There is much talk in this book of the exothermic insects – those that are cold-blooded, unable to produce their own body heat, which is why they generally need warm conditions to function. It may surprise you, then, to learn that the Arctic is home to a surprising number of fly species; among the hardiest of these being non-biting midges. In a place with an average annual temperature of –40°C (–40°F) these tiny flies have adapted in remarkable ways to survive the most hostile conditions. *Smittia velutina* is one such species that is an important pollinator of Arctic plants. It is believed to reproduce by parthenogenesis, which would certainly be easier than trying to find a mate in sub-zero temperatures when nectaring is far more of a priority within the tiny windows of opportunity permitted by seasonal conditions. The sparse flowers have other uses too: the midges shelter under them during rainfall, and traverse around the flower heads to maximize their exposure to the weak warmth of the Arctic sun.

Alderfly

Sialis lutaria

Megaloptera / Sialidae

These attractive insects, with cut-glass wings, spend most of their lives as larvae and live about a year altogether. The eggs are laid in large, rather neat batches on the underside of leaves. Unusually, several females will contribute to the same egg mass. Once the larvae hatch, they must drop from the leaf down into the water, a long way when you are just a couple of millimetres long.

Green Tiger Beetle

Cicindela campestris

Coleoptera / Carabidae

Like its feline namesake, the green tiger beetle is equipped with speed, agility, stealth and huge, powerful jaws that makes it a top predator in its habitat. Half leaping, half flying, its long legs can cover the ground extremely quickly as it chases down and grabs smaller invertebrates. The stunning beetle has a metallic, viridian exoskeleton that glints like an emerald on the sunlit heathland.

Cinnamon Sedge

Limnephilus lunatus | Trichoptera / Limnephilidae

Above: Caddisflies look like moths, but don't have the wing-scales.

Opposite: Chan's megastick belongs to the genus comprising the longest insects ever described.

This is a species of caddisfly, a group of insects related to butterflies, but with wings covered with short hairs rather than scales. They are also mainly confined to the compass of freshwater, with aquatic larvae. The larvae are celebrated for their very unusual body protection. They don't just wander around the surface naked, but instead cover themselves with something resembling a rudimentary sleeping bag, all-over protection made from silk spun by the larvae and with fragments from the environment adhering to it, such as sticks, leaves and sand. It's an aquatically draughty sleeping bag, with the head, tail and legs protruding, but it's great for camouflage. Each caddisfly larva produces a case with a species-specific imprint. The name 'caddis' refers to a strip of cloth.

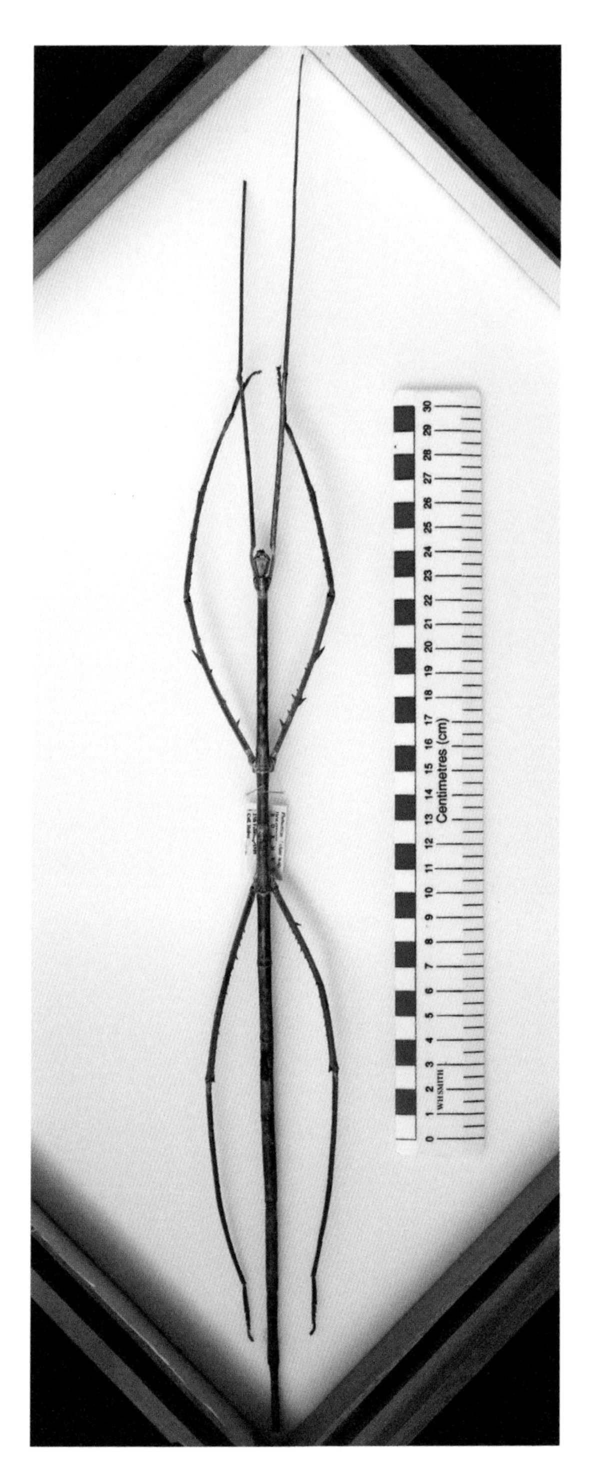

CHAN'S MEGASTICK

Phobaeticus chani

Phasmatodea / Phasmatidae

The insect with the longest body (currently described) is Chan's megastick, which lives high up in the rainforest canopies of Borneo. It has a body length of around 37cm (14½in); its legs take it to around 56cm (22in) overall. This giant stick insect was officially described to science and confirmed as the longest insect in the world in 2008, and has remained in the top spot ever since. Such a colossus should be fairly conspicuous, however, Chan's megastick has only been seen a handful of times. It eggs are unlike those of most other stick insects, in that they have wing-like appendages, which are thought to maximize wind dispersal from the top of the canopy.

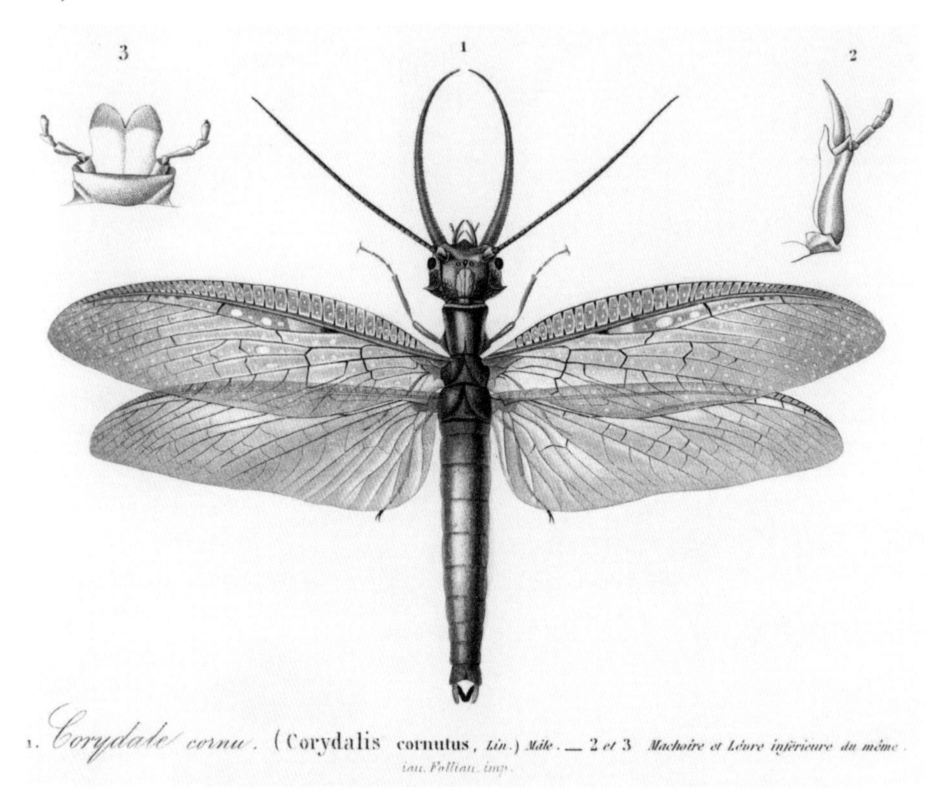

EASTERN DOBSONFLY

Corydalus cornutus | Megaloptera / Corydalidae

The male dobsonfly's huge mandibles are apparently used only to tickle the female.

You've got to love an insect whose larvae are called 'hellgrammites'. And sure enough, these are dominant predators up to 6.5cm (2½in) long, living in streams and eating anything invertebrate that moves. You've also got to love an insect in which the male has huge mandibles but doesn't eat, apparently using them only to stimulate the female. The female, inevitably, does bite, even though its jaws are smaller. You've also got to love this insect because it is huge, with a wingspan up to 18cm (7in), and it flies around lights in midsummer. What an awesome beast.

Nobody knows why they are called 'dobsonflies'.

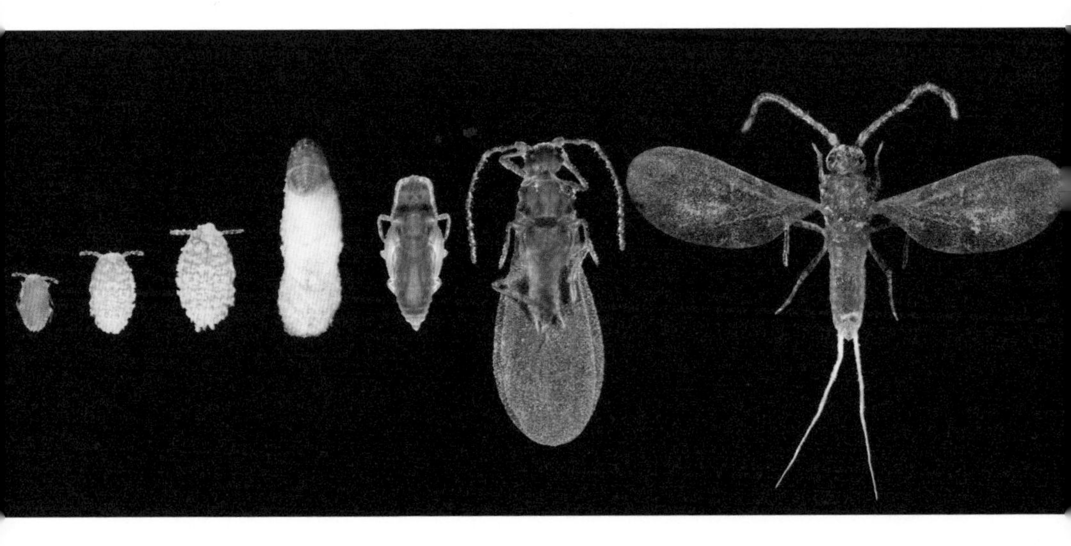

Tamarisk Manna Scale Insect

Trabutina mannipara **| Hemiptera / Pseudococcidae**

Scale insects have their very own complex life journey.

Scale insects are small, obscure and largely unloved insects related to aphids, which suck the sap of plants and void the excess as a sickly, sweet substance through their rear end. Many are considered as pests. But one scale insect may be a character in one of the most famous stories in the world. It is mentioned in both the Bible and the Qur'an that, during the Exodus from Egypt, the Israelites subsisted on a mysterious substance called 'manna', originating from Heaven (hence the phrase 'manna from Heaven'). The description of the substance fits the exudate from scale insects and there is even a candidate, the tamarisk manna scale, which occurs in the right area. Even today, this modern-day manna is sold for human consumption, having been collected in the wild.

Glanville Fritillary

Melitaea cinxia | **Lepidoptera / Nymphalidae**

The stunning Glanville fritillary is a fitting memorial to an intriguing life.

Eleanor Goodricke was born in Yorkshire, England, in 1654. In her youth she developed a passion for butterfly collecting. After being widowed early, she remarried a certain Richard Glanville who abused Eleanor and the couple separated in 1798. Eleanor threw herself into her entomological passion, but upon her death in 1709, the disenfranchised Glanville family contested her will, mainly on the grounds that no sane person would beat bushes for larvae or pay people to provide her with such evidently absurd things such as butterflies. The court wrongly declared that Eleanor Glanville was insane. The family, with their newfound wealth, drifted into obscurity. Eleanor Glanville, on the other hand, is remembered today. One fine day during her troubled life, one of Europe's most intoxicatingly beautiful butterflies found its way into her net, and became the type specimen now recognized throughout the world in her honour.

Elm Zig-zag Sawfly

Aproceros leucopoda | Hymenoptera / Argidae

This tiny elm zig-zag sawfly larva has no idea what a pretty pattern it is creating as it eats.

On the edge of an elm leaf, in the middle of a tree, a tiny caterpillar begins to eat. It starts to chew the margin and heads into the blade, carving a path through the green leaf tissue. Its sole purpose is to consume enough energy to pupate, and so it scarcely stops for a break, just eating whatever is in front of it. What this creative larva doesn't see is the funky pattern it is making in the leaf; zoom out and you will see how the elm zig-zag sawfly gets its name. From the top the caterpillar has carved out a beautiful, meandering zig-zag pattern in the leaf. When the larva has eaten its fill, it spins a delicate, filigree cocoon in which to pupate, emerging around a week later as a smart, black adult sawfly.

Arctic Bumblebee

Bombus polaris **| Hymenoptera / Apidae**

Even summer in the Arctic is cold, so this bumblebee has a particularly fuzzy coat to keep it warm.

Insects inhabit every biome on Earth, but the Arctic presents a particular challenge to small, cold-blooded creatures. In this hostile environment, where temperatures in the short summer season barely exceed 5°C (41°F), the Arctic bumblebee has adapted to conditions. Like all bumblebees, its chitinous exoskeleton has modified into the fuzzy pile of fine 'hairs', known as setae, but in Arctic bumblebees this fur is particularly dense, shielding them from the harsh temperatures. The queen can also pump a constant supply of 30°C (86°F) haemolymph (the insect version of blood), from her thorax to her abdomen to keep herself warm, and maintain her eggs at a viable temperature, allowing this hardy bee to occupy the most unlikely ecological niche.

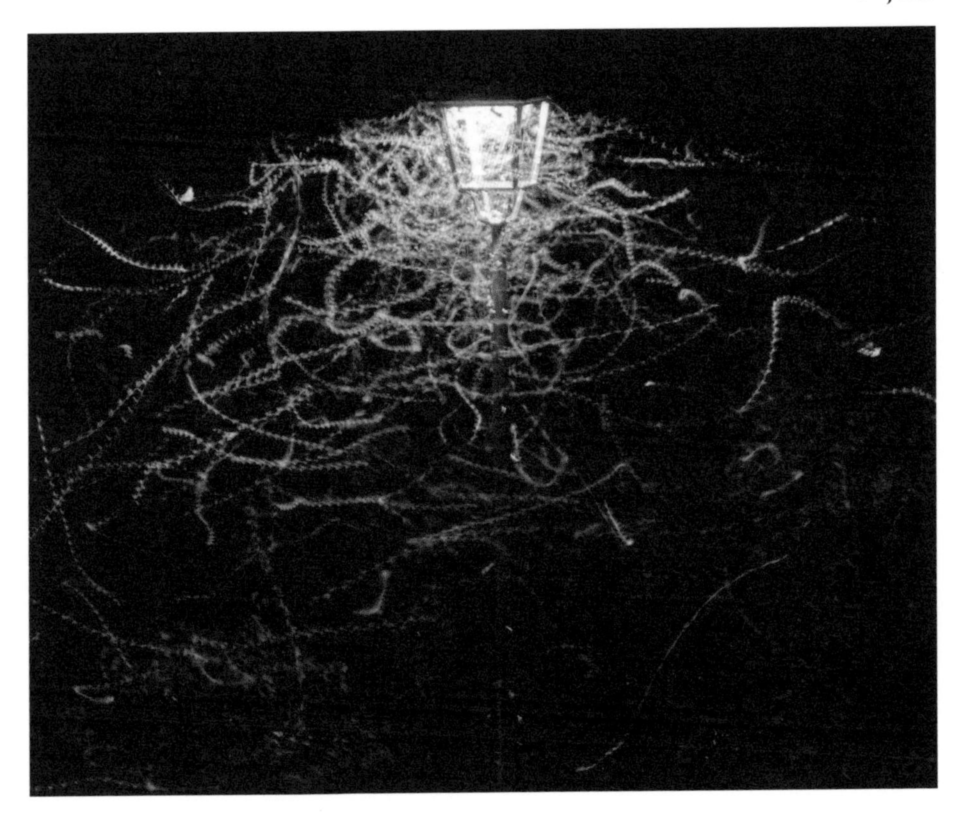

A Moth to a Flame

Lepidoptera

The proverbial insect confusion – insects have evolved to keep their backs to the light, but artificial light disrupts this.

i was talking to a moth
the other evening
he was trying to break into
an electric light bulb
and fry himself in the wires
…
but at the same time i wish
there was something I wanted
as badly as he wanted to fry himself

From *Archy and Mehitabel* by Don Marquis (1927)

ANTLION

Neuroptera / Myrmeleontidae

If you're a tiny insect trying to survive in a sandy habitat, this antlion larva is the stuff of nightmares.

Let us enter the Star Wars universe. In the desert sands of Tatooine lies the Great Pit of Carkoon, and at the bottom of that pit lurks the Sarlacc, a monstrous creature in possession of a gaping radial mouth lined with rings of razor-sharp teeth. Anyone – or anything – unfortunate to fall into the sandy pit will cascade down towards the terrifying maw, and attempts to climb up the loose sand, out of the pit to safety are fruitless. A family of large lacewings, called 'Antlions' could have been the inspiration for the Sarlacc, because their larvae live in exactly the same way. Antlion larvae make a pit in the sand and lurk in wait at the bottom. Anything that enters the pit is swiftly grabbed by the antlion's huge, caliper-like pincers and dragged into the sand. It's a marvellous example of art imitating life, not the other way round...

Common Glow-worm

Lampyris noctiluca | **Coleoptera / Lampyridae**

This soft, green pinpoint of light is a siren call to male glow worms, who can see the signal from some distance away.

There are several types of invertebrates which are commonly known as glow-worms; this one specifically is a beetle. Females are wingless, unable to travel far to find a mate, so instead they use bioluminescence as a beacon to advertise their availability to males, as is recounted in this delightful poem:

...All night she signals him in:
come find me – it is time – and almost dawn;
all night he looks for her in petrol stations,
villages and homesteads, the city's neon signs:
where are you – it is time – and almost dawn...

'Love Poem, Lampyridae (Glow-worms)' by Fiona Benson (2019)

Meadow Katydid

Orchelimum gladiator **| Orthoptera / Tettigoniidae**

Above: Ready for battle, the meadow katydid is a ruthless competitor.

Opposite top: A tiny, furry hulk, the provence hairstreak is a delightful spectacle in summer.

Opposite bottom: The striped shieldbug is truly unmistakeable.

It's singing season down in the meadows of northern North America, where the katydids stridulate their stuff by rubbing their wings together. In this katydid species, the singing males space themselves out just 1.7m (5½ft) apart, each playing their brief tick-and-buzz repeatedly to create a communal sound, so that females are attracted by the beguiling commotion.

Not every male is a welcome member of this choir, however. Every so often, a singer leaves its song post and violently attacks a neighbour, using both its mandibles and claws to inflict genuine damage if it can. A skirmish may prove decisive, or can persist, but eventually there is a loser who is banished to the periphery of the singing group (the lek). Once there, he is unlikely to encounter a female, let alone mate with one.

Provence Hairstreak

Tomares ballus

Lepidoptera / Lycaenidae

In a lepidopteran parallel universe, a Small Copper Butterfly is exposed to gamma radiation and transforms into the Hulk, but with the added bonus of fur. The provence hairstreak appears to be stuck mid-mutation – half elegant butterfly, half hairy green bear, and this is what makes it so endearing. This tiny Lycaenid, found in southern France, Iberia and north Africa, is reputed to be the hairiest butterfly in Europe.

6TH JULY

STRIPED SHIELDBUG

Graphosoma italicum / Pentatomidae

Shieldbugs can be pretty fancy, and this one is no exception. The striped shieldbug, with its strong vertical red and black lines, looks like it has walked straight out of a sweet factory. Its usual hangouts are bright green hemlock and hogweed, so this contrasting livery is most likely aposematic – a warning colouration of potential toxicity.

Common Green Lacewing

Chrysoperla carnea **| Neuroptera / Chrysopidae**

Green lacewing adults are delicate and harmless, but their larvae are the stuff of nightmares.

Not only is the common green lacewing the snazziest shade of green, it also has hypnotic, opalesque eyes and exquisite, lead-lined windows for wings. The female ensures that her young have the safest start possible; eggs are laid on a blob of sticky mucus, which is stretched into a long, thin thread and fixed to the underside of a leaf or to stem. The mucus filament hardens, leaving the egg suspended, seemingly in mid-air like a cotton bud (or Q-tip, depending on your global location). What hatches from the eggs is an aphid's worst nightmare. The green lacewing larvae is an eating machine with huge, curved, pincer-like mouthparts that make short work of small, soft-bodied invertebrates. It covers itself with aphid exuviae (the 'skin' shed from moulting), plant matter and the desiccated bodies of its prey to disguise its smell and appearance, tricking the ants that guard aphid colonies and gaining access to its prey like a murderous Trojan horse.

SPECKLED BUSH-CRICKET

Leptophyes punctatissima | **Orthoptera / Tettigoniidae**

The tympanum (a thinly membraned hole in the 'knee' joint) is clearly visible in this image, and is a common feature in bush-crickets.

The quintessential sound of summer is the chirrup of grasshoppers and bush-crickets. These stridulations are courtship calls – messages passed across the meadow to advertise their availability to prospective partners. Hearing these transmissions is obviously as important as sending them, but how do you hear with no 'ears'? Speckled Bush-crickets, like all insects, do not have ears on the side of their heads, like humans do. Instead, their hearing organ (a wafer-thin membrane called a tympanum) is located, improbably, in the front legs, next to the femoral joint – the 'knee', if you like. This membrane picks up the vibrations of potential partners and sends them via receptors through the bush-cricket's central nervous system, where they are decrypted. Males call by rubbing their wings together and wait for females to respond. Speckled Bush-crickets stridulations are very soft, and do not carry very far; they are virtually imperceptible to human ears but can be picked up by a bat detector.

QUEEN ALEXANDRA'S BIRDWING

Ornithoptera alexandrae | Lepidoptera / Papilionidae

This species is no less than the world's largest butterfly.

This is the world's largest butterfly, although it pips the goliath birdwing (*Ornithoptera goliath*) by just a few millimetres. Both species occur in New Guinea, the Queen Alexandra's birdwing only in a confined area of rainforest about 100 sq. km (38½ sq. miles) in extent. The females are larger than the males, but both sexes dwarf some of the birds that share their forest habitat. Being the biggest of all butterflies is something of a curse, with specimens attracting large sums from collectors and a black market to go with it.

Ruby-tailed Wasp

Chrysis spp. | **Hymenoptera / Chrysididae**

Despite their stunning, exotic colours, ruby-tailed wasps are thought to be surprisingly well camouflaged within the nests of their hosts.

You may have caught a glimpse of this tiny jewel on a sunny day; a winged gem that glints in the sunlight on a heathland walk, or even as you sit in your garden. These small wasps are known also as 'cuckoo' wasps. They search for the nest holes of digger wasps and, when and entrance is located, the Ruby-tailed Wasp makes a tentative check into the tunnel. When satisfied that the coast is clear, it quickly sneaks in, locates the digger wasp's egg and deposits its own next to it (which is how it gets its 'cuckoo' name). The cuckoo wasp egg hatches and helps itself to the contents of its surrogate environment, including the host larvae. The jewel wasp has evolved some cunning methods of disguise to avoid detection by the host. Its brightly metallic body is thought to be difficult to detect on the host wasp's visual spectrum, and the interloping ovum even smells like the one it has supplanted, which ensures that it is not sniffed out by the proprietor.

Giant Mayfly

Hexagenia limbata **| Ephemeroptera / Ephemeridae**

The adult mayfly has a short, but very exciting life.

There should be a firework called a mayfly. It's an insect that explodes for a brief moment and then fizzles out. For all but two days of the year, the giant mayfly lives an obscure life buried the mud in rivers and lakes. Then, often in July, all the adults hatch, embark on a mating dance and pair up. The female lays up to 8,000 eggs, all within a maximum of 48 hours.

These mayflies are most common around the Great Lakes of the USA and Canada. On a big hatching day, millions can be attracted to lights and at these times they may get all over cars, properties and people. On roads, million upon million may be crushed into a messy pulp and make the roads hazardous. At times, authorities have had to use snowploughs to clean them up.

Bee-wolf

Philanthus triangulum **| Hymenoptera / Crabronidae**

Female European bee-wolves have all the tools for great parenting: supreme hunting skills and a built-in medicine cabinet.

The female bee-wolf is a hunter of honey bees. Her method is to ambush an unsuspecting victim as it forages for pollen and nectar, inject it with a paralytic, then gather it up in her legs and airlift it off to her nest. This macabre double-decker is an impressive sight – transporting a live honey bee the same size and weight as yourself is an astonishing feat. She has also evolved a parental superpower to pass on to her larvae; she is an anti-microbial dispensary. She produces a white ointment containing *Streptomyce* – a symbiotic bacterium that counteracts fungal growth – and squeezes it from, of all places, her antennae, by way of special glands which are positioned by the gaps between her antennal segments. She anoints the walls of the nest cell and the eggs, creating a barrier that prevents moisture and pathogens from damaging the egg and larvae. The potency of this 'magic lotion' is now being researched, as its pharmaceutical applications could potentially be extremely beneficial for humans.

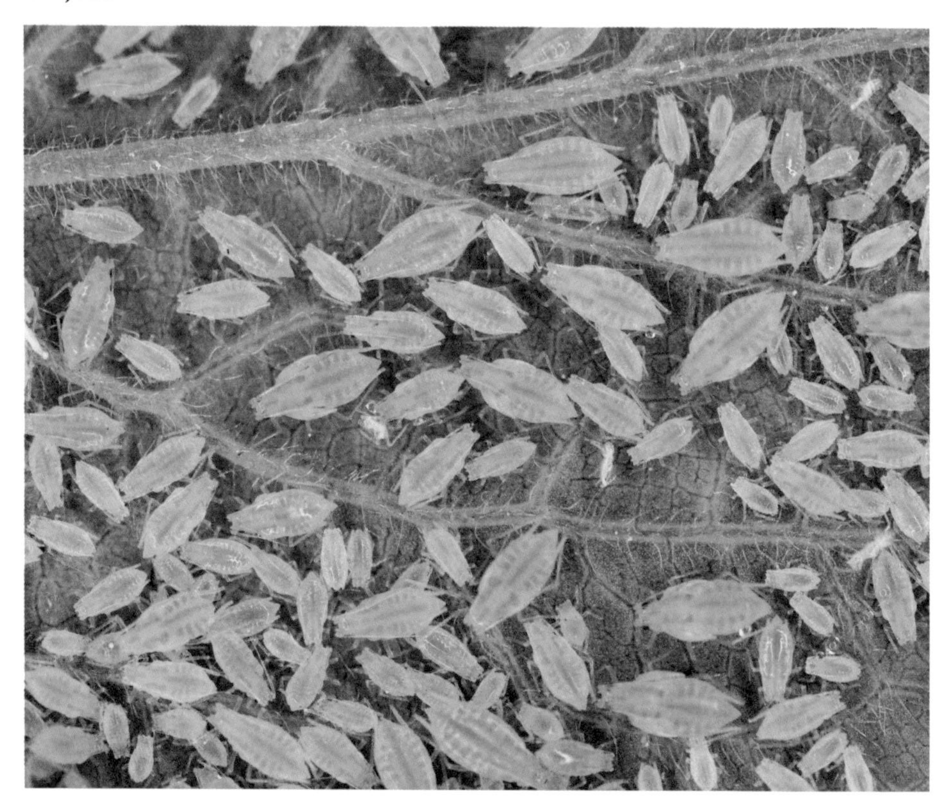

Plum-reed Aphid

Hyalopterus pruni **| Hemiptera / Aphididae**

Plum-reed Aphids have two food plants in completely different habitats.

Aphids are the familiar diminutive bugs that suck sap from many plants. The plum-reed aphid is an example of one with a 'dioecious' lifestyle – it has two host plants, changing from one to the other in mid-season. In this case, eggs overwinter on plum and other trees and hatch in spring, lasting for up to 13 generations, then switch to their secondary host, common reed in midsummer before returning to plums in the autumn. The reason they do this is that plum tree sap is full of amino acids for the plant's spring flush of growth, then sap nutrients decline sharply. Meanwhile, common reed grows and transports enough nitrogen in the summer for the aphids to thrive, but then dies back in autumn, so the aphids return, conveniently coinciding with late season flush of nitrogen on their original host.

Lacebug

Hemiptera / Tingidae

Lacebugs are barely visible to the naked eye, but closer inspection reveals a marvellously intricate body structure.

The tiny lacebugs (2–10mm/$^{1}/_{12}$–$^{2}/_{5}$in) are a delightful group of insects that has spared no expense when it comes to personal decoration. The veins of the forewings are delicately and exquisitely arranged like the finest Torchon lace, and in many species, there are identical plate-like appendages that stick out from both sides of the pronotum, extending the pattern further. Some are transparent, like glass, and some have modified wing shapes with brown or black markings that alters their entire shape and makes them look much more substantial than they really are. Lacebugs are fairly host-specific, meaning that they tend to stick to a single host plant species. These can include gorse, thistle, wild thyme and rhododendron and even moss, where they hide in nooks in the vegetation.

EUROPEAN PRAYING MANTIS

Mantis religiosa **| Mantodea / Mantidae**

As a male Mantis, this might be the last thing you ever see.

This remarkable insect is known for a number of things, but foremost among them is a gruesome habit known as sexual cannibalism. The female, which is by far the larger, routinely eats the male during the course of copulation. This can happen before the act, during or afterwards. In reality sexual cannibalism is by no means the definitive end result, and many males escape to copulate another day, and often do, with only about 30 per cent of encounters in the wild proving fatal. The strange triangular head and goggle eyes additionally give the mantis a splendidly alien look. As if the praying mantis wasn't gruesome enough, it's recently been discovered that it sometimes eats young birds in the nest.

Burying Beetles

Coleoptera / Silphidae

Grave diggers – if it wasn't for the activities of burying beetles and their colleagues, we'd be knee-deep in corpses.

In a book about insects, sometimes you have to hear it for the entomologists, who study them. One of their famous true stories deserves the retelling. Many many years ago, in high summer, two beetle enthusiasts travelled to the New Forest, in southern Britain, a famously insect-studded location. In their perambulations, they came across a human body, a man who had been sleeping rough and succumbed.

It was a tragedy, and the police needed to be contacted urgently. It was also, though, an exceptional opportunity. The two collectors took out a sheet and, with one taking the head of the corpse and the other the legs, manoeuvred the unfortunate deceased over the sheet and shook him vigorously. As they had hoped, out fell a veritable cornucopia of rare and unusual burying beetles. Joyfully, the collectors filled their boots, and to this day, many of their specimens are still kept in the UK's world-famous Natural History Museum.

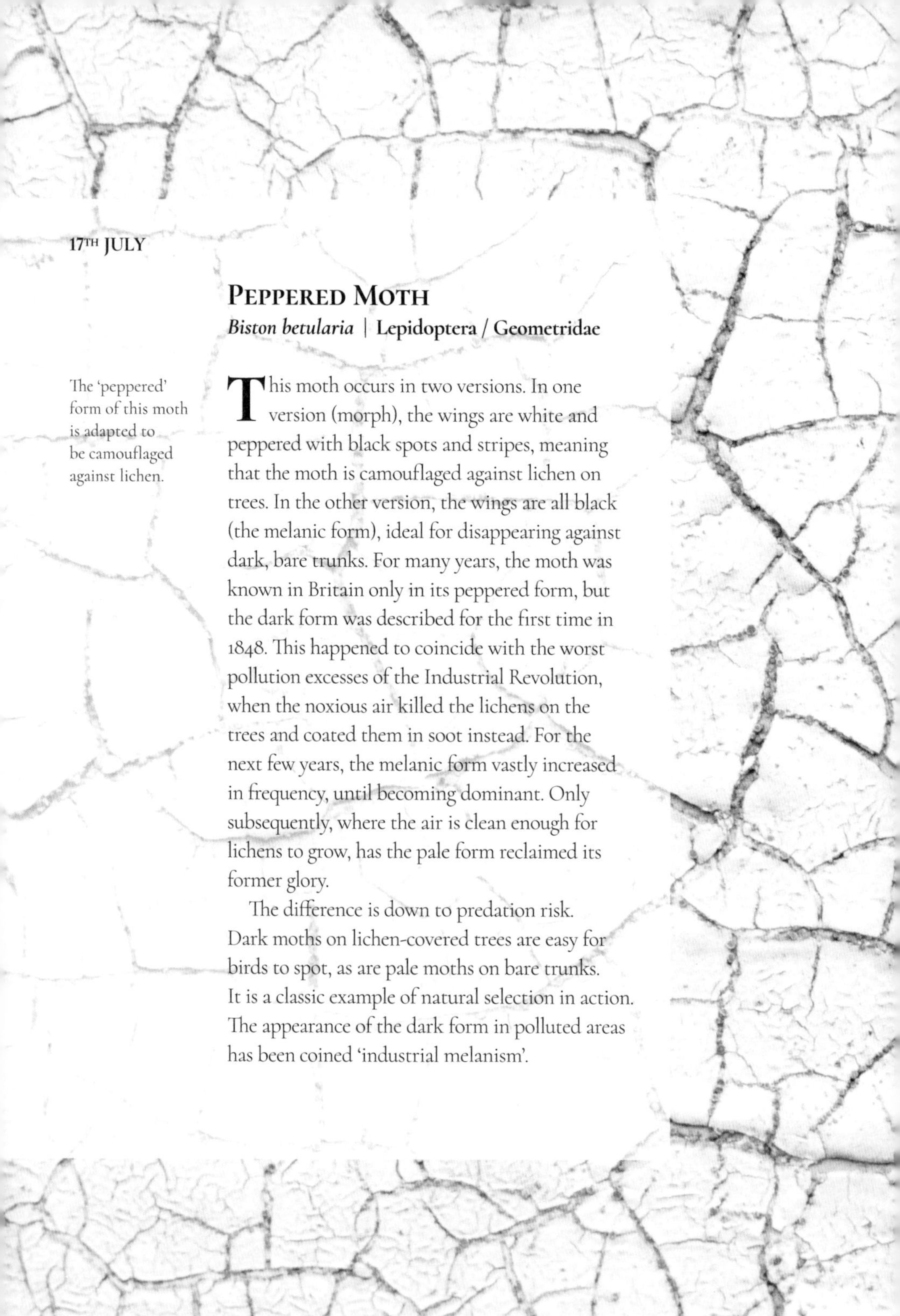

Peppered Moth

Biston betularia **| Lepidoptera / Geometridae**

The 'peppered' form of this moth is adapted to be camouflaged against lichen.

This moth occurs in two versions. In one version (morph), the wings are white and peppered with black spots and stripes, meaning that the moth is camouflaged against lichen on trees. In the other version, the wings are all black (the melanic form), ideal for disappearing against dark, bare trunks. For many years, the moth was known in Britain only in its peppered form, but the dark form was described for the first time in 1848. This happened to coincide with the worst pollution excesses of the Industrial Revolution, when the noxious air killed the lichens on the trees and coated them in soot instead. For the next few years, the melanic form vastly increased in frequency, until becoming dominant. Only subsequently, where the air is clean enough for lichens to grow, has the pale form reclaimed its former glory.

The difference is down to predation risk. Dark moths on lichen-covered trees are easy for birds to spot, as are pale moths on bare trunks. It is a classic example of natural selection in action. The appearance of the dark form in polluted areas has been coined 'industrial melanism'.

Western Sheep Moth

Hemileuca eglanterina

Lepidoptera / Saturniidae

Moths are among the most beautiful creatures on the planet, but because the majority of them are nocturnal, we (typically) diurnal humans miss out on much of this visual smorgasbord. Luckily, there are some moths which fly by day, including the visually stunning western sheep moth. The bright colours, geometric patterns and thick, kohl outlines would not look out of place in one of Picasso's portraits at the height of his Cubist phase, and it could easily be mistaken for a butterfly as it flits about in the dappled sunlight. It is found in forests along the Pacific coast, and through the western states of North America, where the larvae feed on a variety of trees and shrubs, such as wild rose, cherry, and mountain lilac.

Privet Hawk Moth

Sphinx ligustri **| Lepidoptera / Sphingidae**

Above: Privet hawk moths use the moon to orientate themselves at night.

Left: Wonderfully Cubist colours and black outlines are characteristic of this striking day-flying moth.

How do you find your mate when you're nocturnal? The well-known answer is for the female to send out a scent trail of pheromones to lure enthusiastic males. But is this enough? In a recent experiment on privet hawk-moths, scientists found that males are much more efficient at finding females if they can also use celestial cues to orientate themselves – in this case, the moon. Releasing male moths at a distance of 105m (350ft) from captive females, they found that the suitors' arrival coincided with the height of the moon in the sky. The higher it was, the more quickly they arrived, even if it was covered by cloud.

Banded Digger Bees

***Amegilla* spp. | Hymenoptera / Apidae**

Above: Bright azure eyes and strong abdominal bands signal the unmistakeable presence of banded digger bees in your garden.

Opposite top: People and cave crickets go back a long way.

Opposite bottom: Squeak preview – this beetle makes a loud noise when alarmed.

The furtive buzzing and zipping of the banded digger bees are a spectacle of summer across the Iberian Peninsula and Canary Islands. *Amegilla* sp. are a genus of solitary bee that are commonly seen in the sub-desert climes of southern Europe and the North Africa. With dazzling green-blue eyes and striking abdominal bands, they seek out tubular flowers that suit their long, thick probosces. Females nest in stony crevices, or by tunnelling into sandy soil where, despite their solitary lifestyle, they will often (as with many bee and wasp species) nest communally, creating larger aggregations – possibly a safety-in-numbers strategy, or simply all packing into the most suitable soil conditions. Nest chambers are provisioned with pollen obtained using a method known as 'buzz pollination', in which the bee holds the flower with her tarsi ('feet'), and pumps her wing muscles vigorously, sonically displacing the pollen and shaking it onto her body and legs.

21ST JULY

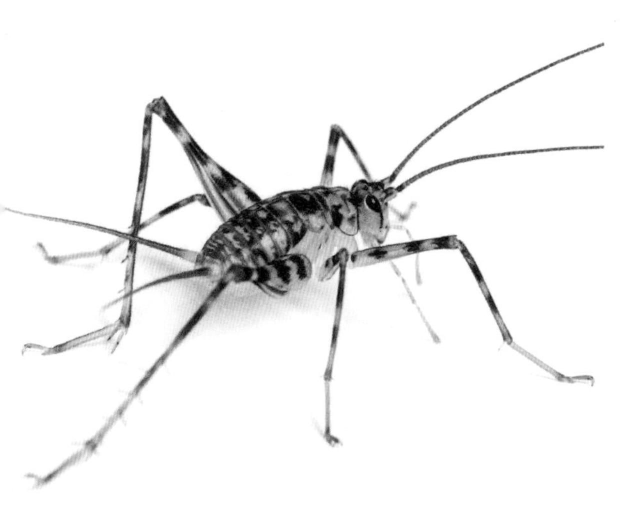

Cave Crickets

Orthoptera / Rhaphidophoridae

What insect would you guess to be the first one ever to be depicted by humankind? Here's a clue – people lived in caves a long time ago and that's where they did their artwork. So, perhaps the answer isn't surprising, that 15,000 years ago, somebody engraved a cave cricket, a fellow speleological, on a fragment of bison bone. This priceless treasure was found at the cave of Les Trois-Frères, near Ariège in South West France, in 1914.

22ND JULY

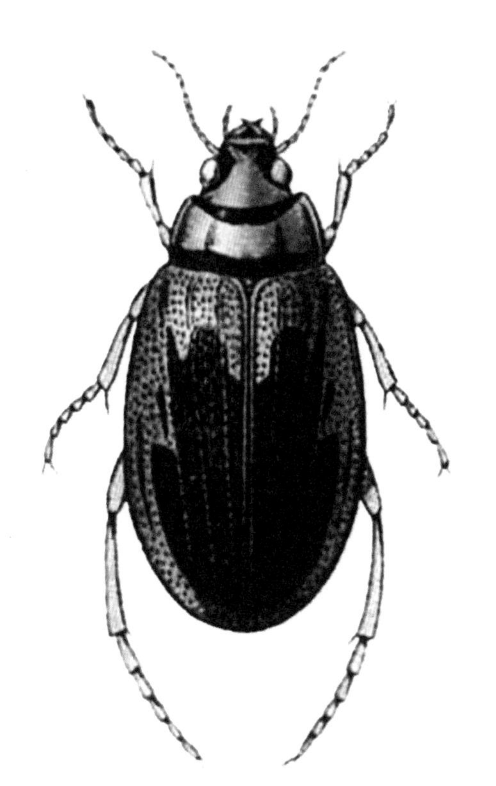

Squeak Beetle

Hygrobia hermanni

Coleoptera / Hygrobiidae

This beetle lives on a bubble and squeaks. In common with other water beetles, it holds an air bubble beneath it when in the water, to allow it to breathe. Should it be threatened in any way, its signature reaction is to make a loud, unexpected noise by rubbing the tip of its abdomen against underside of its elytra.

Alpine Longhorn Beetle

Rosalia alpina **| Coleoptera / Cerambycidae**

Blue is an unusual pigment in beetles, but this longhorn has it in spades.

The lowland alps are home to a very funky longhorn beetle. Blue is an unusual colour in the insect world, and few species pull it off with such aplomb as the dreamy Alpine longhorn. Though not strictly confined to Alpine regions; this rare and protected beetle is distributed widely through the montane regions of central and southern Europe, though it is under increasing pressure as its ancient forest habitat dwindles due to industrialisation. The beetle's elegant colouring comes from a covering of short, dense, ice blue hairs punctuated with rich black. The spectacular antennal segments alternate in blue and black with magnificent tufts on the basal segments. The Alpine longhorn is saproxylic; it spends the larval stage of its life (up to three years) as a pale, pudgy larva chewing its way through dead and decaying wood in ancient woodlands, so it's only fair that it should emerge like this to have its splendid moment of azure glory.

Saharan Silver Ant

Cataglyphis bombycina **| Hymenoptera / Formicidae**

The Saharan silver ant runs very fast in the midday sun.

The Sahara Desert is an inhospitable place, and at midday, with temperatures nudging 50°C (122°F), it is lethal for many animals. For the Saharan silver ant, however, it is an opportunity. Small animals that have succumbed to the heat can be scavenged.

Thus, each day, the silver ants break out from their underground colonies and traverse the broiling dunes on the lookout for the fallen. They are able to check a wide area quickly, because they are among the fastest animals in the world for their size, clocking at 3.1km/h (2mph), equivalent to a person going 720km/h (450mph)! They have special reflective hairs all over their bodies, which reflect the lethal rays, and they produce heat shock proteins before leaving the nest.

They scavenge at temperatures between 47°C (116.6°F) and 53°C (127.4°F). In such heat, they can spend only a maximum of 10 minutes a day outside.

Cinnabar Moth

***Tyria jacobaeae* | Lepidoptera / Erebidae**

The cinnabar moth not only metabolises toxic chemicals from its host plant, it turns the chemical into its own defensive weapon.

The cinnabar moth has taken adaptation to a whole different level. Many insects have evolved to become distasteful to predators; the cinnabar is no exception, and it does this with the help of a plant. Ragwort – the plant in question – has itself evolved a highly effective defence strategy; it contains an alkaloid which is poisonous when eaten in large quantities. However, cinnabar caterpillars have adapted to ingest and metabolize this chemical without consequence. The alkaloid accumulates in their bodies, which in turn makes them toxic to most predators (except a few hardcore species that have also adapted to metabolize the alkaloid). This, combined with their brightly coloured bodies, adds up to a pretty effective survival strategy. They are great fun to spot in large clusters as they demolish ragwort throughout the summer months. The orange stripes of the larvae make way for deep crimson markings on the wings of the commonly seen, day-flying adults.

Lesser Stag Beetle

Dorcus parallelipipedus **| Coleoptera / Lucanidae**

Could this be the quintessential beetle? More common than its larger relative, but every bit as charismatic.

Lesser by name, but not by nature. You may not have heard of this smaller, less showy relative of the European stag beetle, but it is every bit as fascinating and possibly even more endearing. The lesser stag is much more frequently seen and, though still restricted to southern parts of the British Isles it could expand into the northern reaches, if its habitat is preserved. The male of this species don't bear elaborate 'antlers', though both sexes have acutely tapering pincers which, can make them look a little ferocious. But these beetles are anything but scary; they are, if you like, a great beginner's beetle. The lesser stag is very chilled out. It never runs, scuttles or scoots – it ambles, trundles and pootles. I recommend that everyone encourage this beetle into their garden, and you can do that by having undisturbed log piles, and deadwood sunk into the soil, where the larvae can spend several years developing before transforming into its marvellous adult form.

ICHNEUMONS

Hymenoptera / Ichneumonidae

Most ichneumon wasps are parasites of other insects.

There are some areas into which even entomologists fear to tread, and one of these is the study of ichneumon wasps, among the most diverse, but hardest to identify of all insect families. There is a certain irony to this, because the name *ichneumon* is Greek for 'one who follows footsteps.' These wasps are almost all parasitoids of the early stages of other insects or spiders, often the larvae of the former. They commonly use their long ovipositor to inject an egg into the body of a caterpillar, for example. The host is often paralyzed while the egg hatches and the wasp's own larvae eats it from the inside. The reason for the extraordinary diversity is that many ichneumons are host-specific. There are so many, you could perhaps write a book, *An Ichneumon A Day*. Some ichneumons are hyperparasitoids, parasitizing the *in situ* larvae of the parasites.

Violet Carpenter Bee

Xylocopa violacea **| Hymenoptera / Apidae**

You never forget your first violet carpenter bee – just look at those incredible wings.

Your first sighting of the violet carpenter bee is one that may stay with you forever. No quantity of photographs and drawings can truly prepare you for seeing one face to face. Its huge (up to 4cm/1½in in females), polished obsidian body is impressive enough, but the real statement comes from those staggeringly beautiful wings, which appear black but then, with a slight change of angle in the light, explode into radiant amethyst and sapphire. This empress of a bee is common in central and southern Europe, where it feeds on a variety of flowers and eponymously excavates nest chambers in wood, accessible through a single entrance hole.

SAURONA BUTTERFLY

***Saurona aurigera/triangula* | Lepidoptera / Nymphalidae**

The saurona butterfly is named after the evil ruler of Mordor in *Lord of the Rings*.

If you are a fan of the *Lord of the Rings* novels written by J. R. R. Tolkein, you will be delighted to know that many scientists are, too, and there is a butterfly genus named after Sauron, evil ruler of the Land of Mordor.

Dr Blanca Huertas, Curator of Butterflies at London's Natural History Museum, proposed the name on account of the black rings on the specimens' orange wings, which recall the villain's all-seeing eye.

Common Wasp

Vespula vulgaris **| Hymenoptera / Vespidae**

Common wasps may seem vindictive and useless, but they are critical to our ecosystems and deserve more empathy and understanding from us.

The common wasp queen begins nestbuilding in spring, usually underground, or in a tree cavity. Initially, all is calm as the passive queen is preoccupied with building, laying eggs and hunting, far too busy to interact with us. Trouble can start though when the young worker larvae pupate and take over nursery and foraging duties. Workers are genetically programmed to protect the nest swiftly and fiercely against threat, often at the expense of humans who get too close. Further conflict arises in late summer as industrious workers are slowly made redundant and deprived of their natural food (a sugary substance provided by the larvae). Seeking nourishment elsewhere, they gravitate to a sugary pint of lager sat on a beer garden table, or the ice cream cone balanced in a child's fist. The behaviour of hypoglycaemic, geriatric workers has earned them the reputation of mindless assailant, but they are intelligent, protective guardians who simply want to live out their retirement in peace and picnic leftovers.

Apollo Butterfly

Parnassius apollo **| Lepidoptera / Papilionidae**

This alpine butterfly is impressively large and delicately marked with small red 'suns'.

The large, almost monochrome Apollo – named after the Greek god of the sun – is found in alpine and mountain regions of central Europe and Scandinavia, at a higher altitude than many other butterfly species. With its wingspan of up to 90mm (3½in), delicate charcoal markings and blood red eye spots, it is a stunning and unmistakeable butterfly as it flits leisurely around the rocky landscape. Adults will often nectar with wings closed, but will open them if threatened, to reveal the intimidating red false eyes on its hindwings. Upon mating, as with many other butterfly species, the male secretes a plug called a sphragis and inserts it into the female genitalia. This prevents any other males mating with her and ensures the continuation of his genes into the next generation. The female will then depart in search of stonecrop – the larval foodplant – upon which to lay her eggs.

SELF-MEDICATING MOTH

Apantesis incorrupta | **Lepidoptera / Erebidae**

The caterpillars of *Apantesis incorrupta* eat certain plants to make themselves better.

Feeling unwell today? If so, commiserations, but hopefully you are able to reach for the medicine cabinet. There is at least one insect, in caterpillar form, that does the same.

The 'woolly bear' caterpillars of *Apantesis incorrupta* are, like all caterpillars, vulnerable to parasites such as wasps and flies that lay their eggs into their skin. However, this moth has a trick; it can feed from the juices of plants that contain poisonous alkaloids. Normally, a moth will limit how many toxic plants it eats in order to keep its diet balanced. However, if the moth feels poorly, its immune system stimulates it to consume considerably higher quantities of alkaloids. This is a form of self-medication, and it increases the moth's chances of survival.

FIELD CRICKET

Gryllus campestris | Orthoptera / Gryllidae

Field crickets are a sound of summer throughout western Europe, but are declining and are almost extinct in Britain, where the poem was written.

I've seen thy dwelling by the scythe laid bare
And thee in russet garb from bent to bent*
Moping without a song in silence there,
Till grass should bring anew thy home-content,
And leave thee to thyself to sing and wear
The summer through without another care.

FROM 'FIELD CRICKET' BY JOHN CLARE (WRITTEN C. 1832)

* Bent is a kind of grass.

YELLOW-BELLIED BEE ASSASSIN

Apiomerus flaviventris | **Hemiptera / Reduviidae**

Bees beware, this flower-coloured bug could be lurking in wait...

The yellow-bellied bee assassin is a striking bug. Its red, yellow and black may look like a harmless mimic, but make no mistake, this bug is more than capable of looking after itself (the name is a big clue). This bug is a member of a family of insects called assassin bugs, and if you wouldn't want to get in its way if you're on its preferred menu. Many assassin bugs are unfussy in their diet, but this one has a thing for bees. From certain angles its yellow, striped underside, does make it look rather bee-like. It captures prey by ambushing it on flowers, and impaling it with its long, rigid rostrum and injecting it with enzymes that liquify the innards. This essentially turns the prey into a milkshake which the assassin bug sucks up through the straw-like tube. But don't let this give you a bad impression of this tough little bug – it is also an excellent parent. It has discovered a clever way to protect its eggs, by coating them with resin from the brittlebush tree, thus rendering them unpalatable to potential predators.

Net-spinning Caddisfly

***Hydropsyche* spp. | Trichoptera / Hydropsychidae**

Caddisfly larvae are almost all aquatic.

Caddisfly larvae are justly famous for their textile-making skills. Larvae of these butterfly-like insects live in freshwater, and the majority spin silk sacs for themselves, which they usually 'wear' for protection. The net-spinning caddisflies put their silk to a secondary use, however, and fix it to plant stems so that it is strung across like a net. The larvae live inside, but the main function of the net is to trap small edible items that are brought along by the current. The net sieves out algae and invertebrates; amazingly, different species spin nets of different mesh sizes and catch different-sized food particles.

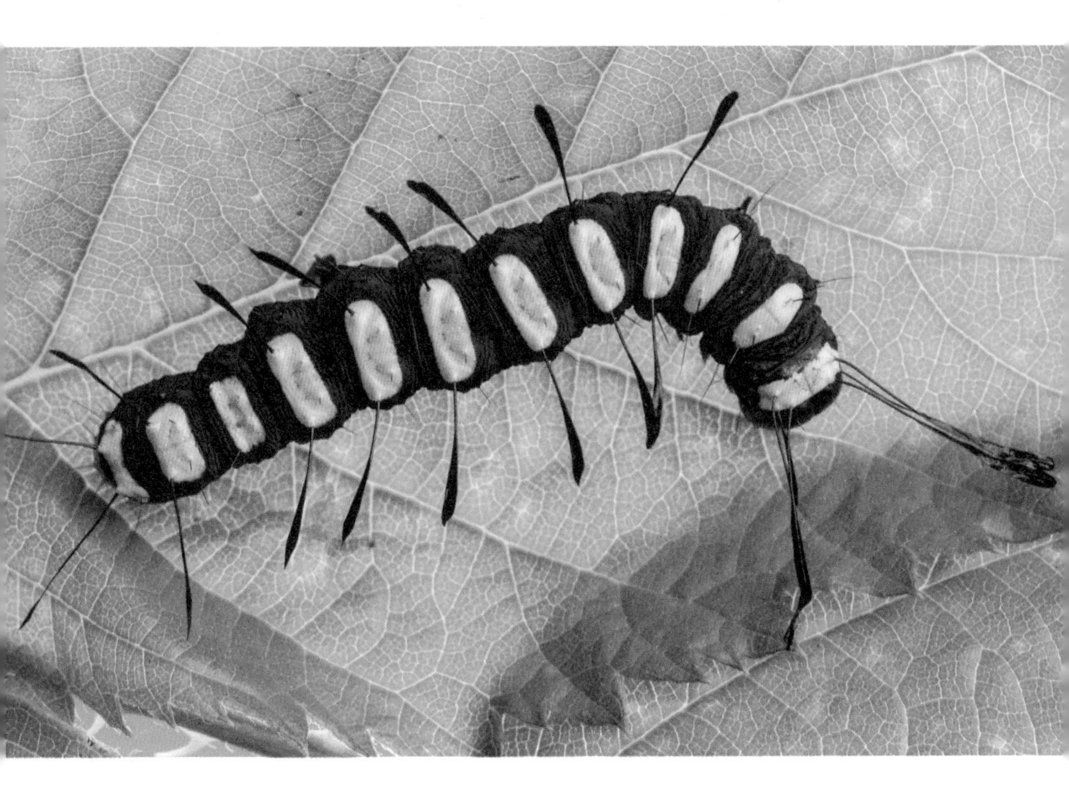

ALDER MOTH

Acronicta alni **| Lepidoptera / Noctuidae**

The alder moth is an expert in reinvention throughout its life.

Reinvention is the *modus operandi* of the alder moth. From egg to imago it transforms its appearance to suit its surroundings. The tiny, newly hatched caterpillar resembles bird poo, which offers it ideal camouflage from predators in the tree canopy. As the caterpillar matures, however, it appears to undertake something of a teenage rebellion. Shedding its faecal form, it turns into a Liquorice Allsort, sporting a matt blue-black body with broad, Day-Glo yellow bands. As if this isn't enough of a departure, it sprouts long, slender, clubbed appendages from its body and looks for all the world as if it is it trying to pick up satellite signals. This bold rebellion is sadly short-lived, as it soon pupates into a subtle, brown moth and once again disappears into its surroundings, though at least with a distant memory of its fluorescent adolescence.

Himalayan Giant Honeybee

***Apis laboriosa* | Hymenoptera / Apidae**

The honey of this large bee has hallucinogenic effects.

Himalayan honey bees are found at high altitude, up to 2500m, throughout the Himalayan range, from northern India through Myanmar, to the mountains of north Vietnam. It is the largest described honey bee in the world, up to 3cm long. Their single-combed nests are built under rocky overhangs or on large tree branches, where they hang like enormous, sugary pendants, weighing up to 60kg. The worker bees forage for pollen and nectar on rhododendron, which is toxic to humans but, in small doses, induces a mildly psychotropic effect. Anyone for hallucinogenic honey? Well, maybe not, thank you, because in larger doses, rhododendron is lethal to humans. Nevertheless, there is a certain demographic of humans who are willing to risk life and limb to harvest the honey for the opportunity to sell it for large amounts of money on the black market, where it is known as 'mad honey'. But please don't try it.

Owlfly

Libelloides macaronius **| Neuroptera / Ascalaphidae**

The owlfly is related to the Lacewings, and hunts insects in flight.

This gorgeous insect catches your eye, and then you immediately think: 'What was that?' It's not quite a dragonfly and not quite a butterfly, but mix the two together and you have an owlfly. This insect is related to the lacewings and antlions, and like the latter it has highly predatory larvae, although they don't build pits (see page 188). The adults, too, immediately recognizable by their clubbed antennae, are predators, catching insects in flight.

Owlflies are noted for their sluggishness. An adult may sit around for hours and hunt for a limited time, in some species as little as 45 minutes a day.

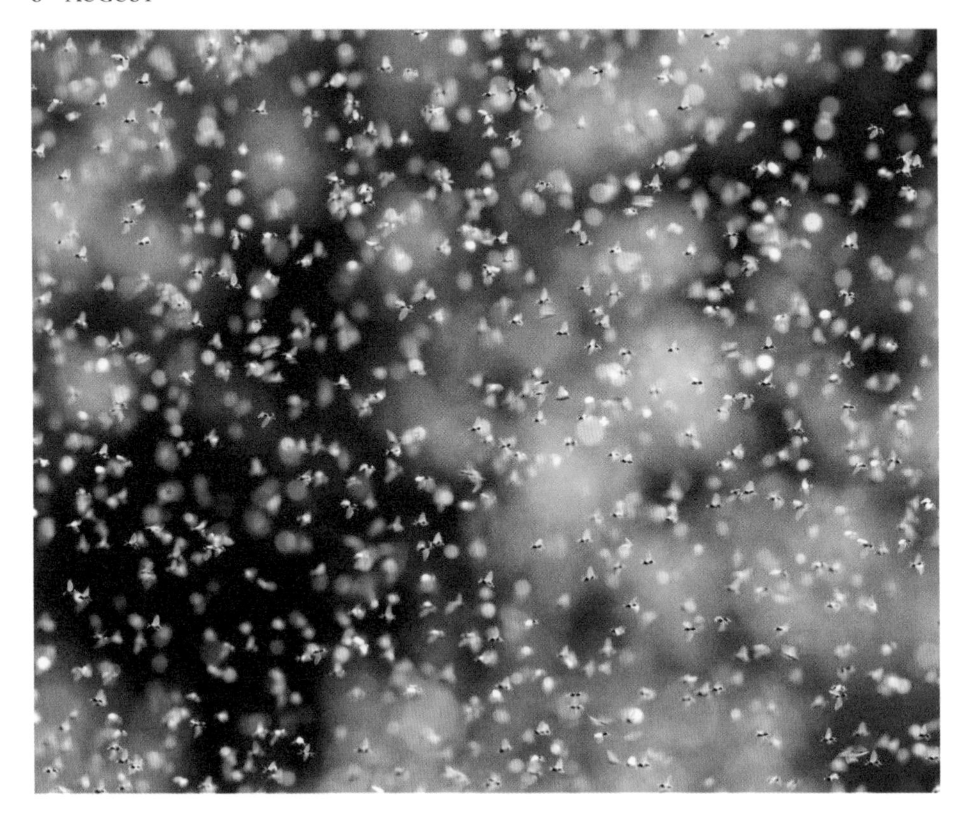

FLYING ANT DAY

***Lasius niger* | Hymenoptera / Formicidae**

The only thing better than the anticipation of flying ant day is the sight of it, as millions of ants take to the air in a nebulous mass.

There is a day in high summer when the air suddenly fills with a million flimsy, winged bodies, attempting to get uplift on the non-existent breeze. These are ant alates – newly emerged queens and males who are setting out on their inaugural flight, and on a single day insect air traffic control in the immediate area is thrown into chaos. It is 'Flying Ant Day', and this winged generation are an evolutionary adaptation driven by a need for genetic diversity. Winged insects can distribute themselves into new habitats and find mates from other colonies, thus widening the gene pool. Once mated, the queens set out in search of a suitable site to form a new colony, losing their wings along the way. The males, however, are unlikely to lose their wings; they simply don't get the chance as their work is over and they will be eaten or die soon after. 'Flying Ant Day' also becomes 'Stuff Your Face' day for many birds, mammals, reptiles and predatory insects.

Giant Scoliid Wasp

Megascolia procer **| Hymenoptera / Scoliidae**

A huge parasitic wasp needs a huge host, and this wasp has it: the colossal rhinoceros beetle.

The parasitic wasps are one of the most diverse groups on Earth, in terms of size and scale. Whereas the smallest known insects, the fairy wasps, are difficult to pick up even with a microscope, the giant scoliid wasp will cover most of your hand. And it needs to be big, because its primary host is another leviathan of the insect world, the rhinoceros beetle. The wasp seeks out the larva of the beetle and lays an egg on it, whereupon the unfortunate beetle larva's days are numbered. Despite its macabre lifestyle, the giant scoliid wasp is exceptionally beautiful; its stocky, black body with yellow markings and slender waist are distinctly waspish, but it is those large, black wings that really steal the show; they are partially iridescent, containing structural layers that refract light in different directions, turning neon blue and green when they catch the light.

Large Dark Olive

Baetis rhodani **| Ephemeroptera / Baetidae**

This species of mayfly can see in the UV spectrum.

If you are a short-lived adult male mayfly, you don't have many tasks to perform. Well actually, just a couple: find a female and mate with her. In order to do the first, some male mayflies have highly specialized eyes, adapted for the purpose of locating flying females in the mating swarm. As well as having 'normal' lateral compound eyes, some mayflies also have a separate pair (turban eyes) that face upwards and are mainly sensitive to ultraviolet light, supremely able to spot the distinctive shape of females flying above them.

Cicada

Cicada orni **| Hemiptera / Cicadidae**

The quintessential sound of Mediterranean summer – the cicada.

There is no mistaking a cicada. The sheer volume of it tells you there is a cicada party in town. The male has a corrugated membrane under its abdomen called a 'tymbal', which it manipulates vigorously (300-400 times per second), producing a sound that can measure up to 120dB. *Cicada orni* is an impressive beast, it can measure 2.5cm in length, with a wingspan of up to 7cm but despite its size, it is not straightforward to spot. It tends to position itself in trees such as pine and olive, too high for us to see easily, a difficulty only compounded by its cryptic, variable colouring that blends beautifully into the tree bark. The trick is to follow the noise slowly to an occupied tree, triangulate it by circumnavigating the trunk, and simply hope it will catch your eye. However, a highly recommended method of interaction is to sit out on a Mediterranean terrace, with a view of the sea and a glass of wine in hand enjoying the serenade.

Dark Giant Horsefly

Tabanus sudeticus | Diptera / Tabanidae

On a hot day, horseflies need a drink, too.

It's a sweltering day in August. Horseflies do something very unusual on hot days like this. They dunk themselves on to the surface of the water in order to drink. This is potentially dangerous for most insects, which prefer to stay by the edge of water rather than risk drowning. But horseflies swoop down on the open water, take a gulp and fly off again, just as a swallow might do.

This odd behaviour is thought to be caused by predation. Despite this species, for example, being Europe's largest hoverfly, which makes an intimidating buzz, it is still vulnerable to predators, including robberflies and wasps. Hence its safety-first drinking habits.

Great Green Bush-cricket

Tettigonia viridissima **| Orthoptera / Tettigoniidae**

The UK's largest bush-cricket is also pretty noisy, drowning out other insects nearby.

Late summer in southern parts of the UK sees the appearance of the largest bush-cricket. The great green bush-cricket is a giant in the scrubby vegetation it inhabits – a stocky, bright green species with huge hind legs, long, spidery antennae and wings that extend beyond the tip of the abdomen. A glimpse of one can be quite surprising, such is its size, and they are marvellously quirky as they totter around haphazardly in low trees and hedgerows. Its high, persistent stridulation is quite loud, and when close by, drowns out other hoppers in the area, however they can be quite elusive and difficult to pinpoint in the vegetation. Females lay their eggs on the ground where they overwinter; the tiny nymphs emerge and develop through late spring and summer. They mature by around late July and can be seen well into autumn.

Red Velvet Ant

Dasymutilla occidentalis **| Hymenoptera / Mutillidae**

If you're foolish enough to ignore the many warning signals of the red velvet ant, then don't be surprised if you have to endure its blindingly painful sting.

The red velvet ant is a delicious concoction of evolution. It's as fluffy as a bee, sprints like a cheetah, and kicks like a mule. It's actually a solitary, parasitoid wasp, though the wingless females are more similar in shape and locomotion to ants. It is a glorious example of aposematism – in which animals develop defence mechanisms to avoid being eaten, and the red velvet ant has quite an array. For starters, it (like other Mutillidae) has an extremely tough exoskeleton and spherical abdomen, both of which make it harder to bite or grab. And if its intense vermilion and black pile isn't a clear enough signal to back off, then the velvet ant will further attempt to discombobulate its assailant, emitting a loud squeak by way of a stridulatory organ on the underside of its body. Should this also fail, it will deploy its last resort – a sting of such blinding intensity as to give rise to its legendary (but thankfully overstated) nickname, 'cow killer'.

BATMAN HOVERFLY

***Myathropa florea* | Diptera / Syrphidae**

The Batman hoverfly has a mark on its thorax which looks remarkably like the Batman logo.

Arguably the best English name of any insect derives from this hoverfly's mark on its thorax, which looks just like the Batman logo! It is a very common European species, and it is a marvellous example of how different larvae can be to adults. The adults are a sun-loving, fast, dynamic species that can beat their wings 120 times a second and pollinate flowers. The larvae are 'rat-tailed' maggots, which dwell among the edible detritus of rancid, putrid water, a habitat so starved of oxygen that the larvae have an elongated end to their bodies that forms a telescopic snorkel to reach the surface. Many of the Batman hoverflies that you see will have begun life in rot holes in trees, where branches have fallen off and left a hole.

Large Blue Butterfly

***Phengaris arion* | Lepidoptera / Lycaenidae**

Above: The large blue has a highly unusual lifestyle for a butterfly which includes phoresy and carnivorism.

Opposite top: The hornet hoverfly doesn't just look like a hornet; it smells and sounds like one too.

Opposite bottom: It's going to be a good day in the Highlands!

If a vegan carrion beetle (see page 251) isn't proof enough that the insect world is a wonderfully cryptic place, then how about a carnivorous butterfly? The large blue looks and behaves in a manner indistinguishable from other butterfly species – that is, until you take a closer look at its larval stage antics, which are extremely uncommon and rarely recorded in butterflies. It begins life conventionally enough, as an egg on wild thyme, where it eats, grows and moults just like any other caterpillar. But once it arrives in its fourth instar, things take an entirely different turn, as it drops onto the soil and secretes a substance which mimics that of an ant larvae. An unwitting passing ant worker transports the disguised interloper back to its nest colony, whereupon the caterpillar completely switches from vegan to carnivore, and begins feasting upon the ant larvae. It will then pupate and overwinter in the nest, emerging the following spring.

Hornet Hoverfly

Volucella zonaria

Diptera / Syrphidae

The hornet hoverfly looks like a hornet, sounds like a hornet and even smells like a hornet, all to simply avoid being eaten by its host. The yellow and brown colouring are perfect subterfuge; once it has secreted itself into the hornet nest, it lays its eggs among the detritus and frass which becomes the food supply for its larvae, a bit like leaving your kids at the world's most dangerous all-you-can-eat buffet.

Highland Midge

Culicoides impunctatus

Diptera / Ceratopogonidae

It's up to 2mm (½in) long and mighty. It affects Scottish tourism and, at certain times of year, dominates conversation. It shows that you don't have to be big to be powerful, especially if you are persistent – and in clouds. The Highland midge is one of the smallest of all blood-sucking flies, common throughout northern Eurasia. But the Scottish people 'own it' more than most and it is certainly abundant there.

Puss Moth

Cerura vinula **| Lepidoptera / Notodontidae**

There can be few things more delightful in life than the evolutionary chaos that is the puss moth caterpillar.

The puss moth gets its name from the fluffy kitten it becomes when it emerges from its cocoon as an adult. Its dense white fur and subtle markings give it a decidedly Burmese countenance. Some might say, however, that it is the puss moth's larval stage that is the true scene-stealer, and it really is a startling looking beast. The back end is fairly standard caterpillar fayre, though it does have two splendid tendrils streaming from its rear. Follow the green, cylindrical body back up towards the head and it turns into an extra from Minecraft: its little round head is encased in a bizarre pink cube, like it's wearing a prawn cocktail crisp box on its head. When threatened it rears up both its boxy bonce and tail streamers, presumably to confuse the heck out of predators and simply bamboozle them into retreat.

Mosquito

***Sabethes tarsopus* | Diptera / Culicidae**

Murder on the dance floor – this mosquito is a killer queen.

'Mosquitoes are my favourite insect!' said virtually nobody, ever. They are an insect with which humans share an extremely problematic relationship worldwide. As sanguivores, female mosquitoes (males do not feed on blood) will readily insert a hypodermic proboscis under the surface of skin to suck out and ingest blood, which triggers the development of their eggs. It also injects anticoagulant saliva that improves blood flow from the host and triggers our histamine systems, leaving us at best with an itchy lump, at worst bacteria, parasites and viruses that cause devastating diseases such as dengue, malaria and Zika. But it is complicated. Mosquitoes are a large component of the global food web, feeding billions of other organisms. The aforementioned anticoagulant saliva has potential pharmaceutical applications that could facilitate major breakthroughs in human-based medicine. And my goodness, some of them look marvellous. *Sabethes torsopa* is one such species that is to be viewed with caution (it is a vector of nasty illnesses) but just check out those flares; the hindlegs are adorned with long metallic fringes. Often, just males will have the great outfits as they need to show off for mating rights, but in this species females have them too, and groups of them together form the greatest ABBA tribute act in the insect world.

Golden Tortoise Beetle

***Charidotella* spp. | Coleoptera / Chrysomelidae**

Possessing pigment as well as structural colour, this beetle has the best of both worlds.

While iridescence is a common phenomenon in the insect world, gold is somewhat rarer and the golden tortoise beetle suitably lives up to its name. The tortoise beetles are an unusual bunch, in that their carapace extends outwards into a round shield that conceals the head and legs. Reminiscent of a tortoise shell, and not dissimilar to a 15th-century concept for an armoured tank design by Leonardo da Vinci, this armour provides the beetles with enhanced protection from predators. The peripheral edges of the shield are transparent, but the central area is highly reflective, metallic gold. While its gilding is astonishing enough, the golden tortoise beetle also has the ability to actively change colour. Its exoskeleton possesses both structural colour and pigment layers, and by 'fading out' the structural metallic layers, it can switch from shiny gold to dull, matt red in reaction to external stimuli, in less than 2 minutes.

European Cranefly

Tipula spp. **| Diptera / Tipulidae**

Craneflies are famous for their ability to go about their business despite dropping limbs.

Get a bunch of elite human engineers to design the perfect insect – and they wouldn't come up with the cranefly. Many of us are familiar with these fabulous insects, which seem to embody life's everyday faux pas and disgraces as they bumble along, propelled by apparently inadequate wings and dangly, gangling legs. They are fatally attracted to artificial light on mild evenings and find themselves indoors, flying against windows and accidently shedding legs. They have the last laugh, though. They only need four legs, so why not lose a few? Craneflies have been flying around for 140 million years with great success, so they don't need any humans to pronounce upon them.

Pantaloon Bee

Dasypoda hirtipes | **Hymenoptera / Melittidae**

Female pantaloon bees have amazing, multipurpose 'trousers'.

Summer meadow flowers are the place to find a whole host of different bee species of all shapes and sizes, and one of the most entertaining among them there is the pantaloon bee. It has acquired this name due to its swashbuckling flares. The hind tibiae of females are adorned with pollen baskets made up gloriously long, dense hairs. When fresh, the hairs are bright yellow, and when loaded full of pollen they positively glow. Pantaloon bees nest in the ground in sandy soil, and they are a joy to watch in their excavations; they dig into the sand with their front legs, while their rear legs work in a breaststroke motion to scoop and shovel it backwards and away from the burrow. The females work industriously to create tunnels in which they will build small side cells, each of which they will provision with pollen and deposit an egg that will develop over winter and emerge the following summer.

Rosy Maple Moth

Dryocampa rubicunda | Lepidoptera / Saturniidae

Cake, anyone?

We often refer to highly visually appealing things as 'looking good enough to eat', and in the case of the rosy maple moth this could certainly be the case. From its humble beginnings as a modest green caterpillar, blending in among the leaves, it pupates into a candy-coated vision of patisserie window pastel colours. It is a living Battenberg cake – its lemon yellow body and wings, swirled with carnation pink. The rosy maple moth is native to North America, where the larva bumbles around in the canopy, eating the leaves of its food plant, the maple tree.

Zombie Ladybird Wasp

Dinocampus coccinellae **| Hymenoptera / Braconidae**

The ladybird 'standing guard' over the cocoon is paralysed.

When is a ladybird not a ladybird? When it is a zombie ladybird, of course. That is what this most popular of beetles becomes when it is parasitized by *Dinocampus coccinellae,* the latter thereby becoming pretty unpopular with everyone. The wasp lays its eggs inside the ladybird's body and after a month, with the ladybird paralyzed but still alive, it exits its host's body and weaves a cocoon attached to the ladybird's leg. The beetle thus sits over its parasite's cocoon as if incubating it, and if a predator approaches, amazingly, the paralyzed and barely moving host twitches enough to repel it. The ladybird thus acts as the host's 'bodyguard' in the cocoon.

Remarkably, some wasps are themselves parasitized within the ladybird-guarded cocoon by another type of wasp. This is known as hyperparasitism. Even more amazingly, some ladybirds recover after all this!

Heath Potter Wasp

Eumenes coarctatus **| Hymenoptera / Vespidae**

Potter wasps craft exquisite mud urns for their young.

The heathlands of southern England are home to one of the UK's most endearing and special wasps, which is famous for its craftswaspship. Whereas most wasp species use an existing nook or cranny as a depository for their eggs, the heath potter wasp creates one from scratch. It scouts for a suitably study and sheltered stem in the heather, then collects small balls of wet mud from a nearby puddle or patch of damp ground, which it transports back to the stem. The wasp will then construct a hollow clay pot, moulding the clay into shape with its mandibles, narrowing the neck then giving it a wider lip, similar to an amphora. It then lays a single egg in the pot and goes off hunting for tiny caterpillars, which it brings back and places carefully in the pot with the egg. Once full, the pot is sealed shut with more wet clay, and the egg is left to its own devices, with enough food to see it through to pupation.

27TH AUGUST

Burying Beetle

Nicrophorus orbicollis

Coleoptera / Silphidae

The death of a small animal is the beginning; a personal chapter closes but someone else's opens. The carcass attracts burying beetles, who take away the corpse, bury it and the female lays eggs. This is where the story gets unusual. Both parents, not just one, remain with the eggs and tend the nymphs. Biparental care is rare in the insect world; the male is there to prevent another male finding the carcass.

28TH AUGUST

Blowfly

Diptera / Calliphoridae

Entomology is a common tool in forensic science today. But in 1235, a Chinese judge called Song Ci recalled a time when a farm worker was killed, and his injuries pointed to a sickle as the murder weapon. The judges subsequently summoned the man's fellow harvesters. Their sickles were confiscated. Soon, a significant number of flies gathered on the blade of one particular sickle, attracted by the body fluids. Case closed.

Large Marsh Grasshopper

Stethophyma grossum **| Orthoptera / Acrididae**

Above: Quaking bogs – large areas of wet peatland – are home to these rare, colourful grasshoppers.

Opposite top: Burying beetles seem unlikely candidates to be among the best of all insect parents.

Opposite bottom: Flies are excellent at solving crimes.

The large marsh grasshopper lives a fragile existence in one of the UK's most threatened habitats. Mires – or quaking bogs – were once commonplace in low-lying heathland throughout Britain. However, modern agricultural practices and wetland drainage have reduced these moist, peaty areas to isolated sites that are now protected to try to prevent further loss. Within these fragmented bog remains live specialized animals that rely on the particular conditions provided by this sphagnum moss-rich, acidic habitat, including the large marsh grasshopper. Britain's largest native grasshopper, it lives low down among the vegetation, feeding on grasses and sedges. Females lay their eggs at the base of damp tussocks. It can be identified by the deep vermilion stripe that runs along the lower margin of its powerful jumping legs. It can also, occasionally, take on a pleasing deep pink form, which gives it a distinctly pick 'n' mix appearance.

A Twana Story

Circular Forms. Sun & Moon by Robert Delauney, 1912-13 (oil on canvas).

According to the Twana, a long time ago, there was just darkness. The ants struggled to find food in the dark and were fearful of the grizzly bear Tsimox, who hunted them, digging up their nests and eating their young. The wise Ant Woman realised that if there was light, they would be safer. She visited the the Creator, but the bear followed her and interrupted her plea to the Creator for light, arguing that he wanted to continue living in darkness. The Creator suggested they dance to settle the matter: the winner would get their way. Many came to watch and brought food. The bear ate greedily but the Ant Woman fasted and prayed. The bear would dance, then eat; the Ant Woman would dance, then tighten her belt. This went on for four days, all the while the waist of the Ant Woman became smaller and smaller as she cinched it in further, until the bear, tired and over-full, fell asleep, unable to dance any more. The Creator decreed that both parties should receive their wish, and so the world become light for half the time, and dark for the other, and the ants got their tiny waists.

Leafcutter Bees

***Megachile* spp. | Hymenoptera / Megachilidae**

Who's nicking tiny bits of your garden? Probably leafcutter bees...

If you have noticed small, circular notches appearing in your plant leaves in the summer months then your outdoor space could be home to leafcutters – an intriguing family of bees with an unusual skill. Like many other solitary bees, they nest in cavities in wood, suitably sized hollow plant stems, and masonry. Females find leaves from a variety of plants including roses, willowherbs, fuchsia and honeysuckle, cut circles out of the edges using their mandibles, and then take them back to the nest, where they are pasted to the walls of the cells. Six to fourteen leaves can be used as 'wallpaper', and then the females collect pollen using a bristly patch on the underside of the abdomen, rather than on the hind leg. They twerk energetically across the surface of the flowers as they rub the pollen onto their hairy underside. The pollen is then transported back to the nest cell and deposited, with an egg, and the nest cell sealed up with further leaf segments.

Rainbow Grasshopper

***Dactylotum bicolor* | Orthoptera / Acrididae**

This flightless grasshopper protects itself with bright colours to deter predators.

The gorgeous rainbow grasshopper has got aposematism down to a fine art. Its red, yellow and black spray-paint job are in classic 'don't mess with me' colours, which are even more useful when you consider that this grasshopper is flightless. Indeed, its best escape method is those powerful hind legs that can propel it quickly out of danger. This grasshopper has a large range and is found in deserts and prairies throughout North America, from southern Canada right down to Mexico. The adults are extremely cosmopolitan in their dietary choices – eating a wide range of plants. The nymphs, however, are highly specialized and eat only one species – *Baccharis wrightii*, commonly known as Wright's false willow.

Stylops

***Stylops melittae* | Strepsiptera / Stylopidae**

Stylops are among the weirdest of all insects.

The stylopids are among the strangest insects known, and constitute a small order called Strepsiptera or 'twisted-wing' parasites. The name comes from the free-living male's forewings; the hindwings make them look like they are wearing an outspread cloak. The females are grub-like and live their entire lives within the body of their host, often a solitary bee (*Andrena*). The mobile male, therefore, must follow a pheromone and mate while perched on the host's body!

The female doesn't kill its host, although it does render it infertile. Typically, the front of the stylopid's body partly protrudes through the cuticle of the host, making it look bloated. There is a word for the host's fate – stylopized.

Stylops are just the sort of obscure insects that are a specialist's dream: sure enough, *Stylops melittae* is the emblem of Britain's Royal Entomological Society.

Antler Fly

Protopiophila litigata **| Diptera / Piophilidae**

The rutting season is an impressive spectacle, but did you know there's another tiny rut going on at the same time?

Deer are a familiar sight across farmland, parks and moorlands, though they can be elusive; they are wary of humans and tend to keep a low profile. That is until the rut, when rival stags, pumped with testosterone, square up against one another in noisy, one-on-one brawls. They use their impressive antlers to intimidate and strike each other until one party concedes, the victor earning the right to mate with the does of the herd, thus continuing his lineage. But that isn't all that's going on with those antlers, because zoom right into the velvety pile and you will see an identical battle ensuing. Antler flies do not have antlers (sadly). They live benignly on the antlers of deer (and other suitably-appendaged species). These small flies are remarkably similar to their hosts in that the males also display aggressive behaviour, entering into fierce battles for the honour of mating rights.

Lobster Moth

***Stauropus fagi* | Lepidoptera / Notodontidae**

The lobster moth caterpillar is an evolutionary curiosity; more crustacean than insect.

The lobster moth is a large, attractive moth: fuzzy grey-brown body and warm brown wings with rows of white flecks and what look like, to all intents and purposes, fluffy shoulder pads. None of these characteristics could be described as being remotely similar to a lobster, so why is it named so? Well, the larval stage is very unlike your usual caterpillar. It starts out like an ant which is useful subterfuge but, as it grows, it fills out into a chunky beast with large sections, more crustacean than caterpillar. Its legs are unusually long and spindly, and its rear end bends back over its body and it almost double the thickness. The result of these bizarre morphological adaptations is a caterpillar which looks more like, well, a lobster. In reality its name is down to human projection rather than purposeful adaptation – it has no reason to look like a lobster, but it does make a cracking job of blending in against dead leaves and twigs.

PYGMY GROUNDHOPPER

Criotettix japonicus | Orthoptera / Tetrigidae

The pygmy groundhopper is expert at escaping the mouths of frogs.

Most insects lead short lives with brutal ends. However, if the worst happens and they are captured by a predator, a surprising number of species have one final trick up their sleeves; they feign death. The pygmy groundhopper of Japan, however, plays death a little more cleverly than most. If cornered by its main enemy, a frog called *Rana nigromaculata*, it assumes a death posture that involves spreading its pronotum, hind legs and lateral spines out as wide as possible.

Sadly for the frog, its gape width is limited and it often gives up. It finds the whole thing hard to swallow.

Vegan Carrion Beetle

***Aclypea opaca* | Coleoptera / Silphidae**

Watch out! Unlike its meat-scavenging relatives, this carrion beetle may come for your brassicas.

Carrion beetles eat dead meat. The clue is in the name. They are nature's recyclers, flocking to recently deceased animals – usually mammals – whereupon they will set to work eating through it until, in cooperation with myriad other organisms, there is virtually nothing left. Without the tireless work of these essential decomposers over millions of years, we would be miles high in dead stuff, so we owe a lot to our tiny tidy-uppers. Nature, however, is a funny thing, so what happens when a carrion beetle doesn't eat, well, carrion? It's vegan, of course. *Aclypea opaca* is a species of the Silphidae family of carrion beetles, and one member of a small genus that are phytophagous – both larvae and adults feed on plant, rather than animal material, usually in the form of brassicas and beets.

JAPANESE HONEYBEE

Apis cerana japonica **| Hymenoptera / Apidae**

Japanese honeybees have some extraordinary means of defending their hives.

Honeybee colonies are formidable, but they do have predators that are immune to their stings, such as the Japanese giant hornet *Vespa mandarinia*, which will launch raids involving many wasps. The only defence for the bees is to recognize and neutralize any scouting wasps that take an interest in their hive. If a scout enters, immediately a large number of workers set upon it, surrounding it and making a ball from which it cannot physically escape. They then vibrate their wing muscles to create heat, something that they routinely do in cold conditions. The hornet is now trapped inside a rapidly heating cage of bee bodies. The bees can tolerate temperatures up to 50°C (122°F), but the hornet is vulnerable to anything above 46°C (144.8°F), and furthermore the hard work the bees are putting in raises the level of carbon dioxide. The hornet scout is eventually killed inside its furious ball of captors, dying of overheating and asphyxiation.

Four-barred Knapweed Gall Fly

Urophora quadrifasciata | Diptera / Tephritidae

These striking little flies lay their eggs inside thistle plant tissue.

From midsummer, knapweeds explode into life across fallow fields and meadows, their pinky-purple heads swaying in the warm breeze, and covered in pollinating insects. Knapweeds are insect magnets: the rich nectar is a favourite food source for all manner of bees, flies, beetles, bugs and more, all gorging themselves before the flowers fade into desiccated seedheads. But there is another fly on the knapweed head that is interested in more than just nectar. *Urophora quadrifasciata* is, as its common name suggests, a gall fly. The adults are very striking – their wings are marked with broad, curved black bars that appear to read 'VU'. Females lays eggs into the tissue of the knapweed calyx, just below the surface, pushing them in with a sturdy, black, pointed ovipositor. The eggs hatch and the larvae eat into the internal plant tissue, triggering the formation of a gall around it. There they remain in their snug, safe environment, surrounded by food until they pupate and emerge from the dried seed heads.

MORMON CRICKET

Anabrus simplex **| Orthoptera / Tettigoniidae**

It's coming for you...

This large, flightless katydid spends most of its existence anonymously chewing a variety of plant material in the Great Plains. However, every so often, it suddenly explodes in population and swarms, in a similar way to a locust. Hordes of these animals, which can cover 2km (1¼ miles) a day on foot, devour every plant in their path and, if they catch up with many of the animals in front of them, they devour them too, in unsubtle acts of cannibalism. Swarms of these crickets can cause devastation to crops. The peculiar name comes from a famous outbreak that affected the early Mormon settlers who arrived in what became Salt Lake City, Utah, in 1847. The next year the insects struck. Fortunately, however, gulls from the nearby lake came to the rescue and feasted on so many of the crickets that part of the harvest was saved. The incident became known as the Miracle of the Gulls and there is a Seagull Monument in Salt Lake City.

Diabolical Ironclad Beetle

***Phloeodes diabolicus* | Coleoptera / Zopheridae**

This beetle is no delicate flower – it's modified exoskeleton makes it one of the toughest animals on Earth.

One of the most awful sounds an entomologist will ever hear is the sound of an insect being crushed beneath their own foot. Chitinous exoskeletons are strong and resilient, but everything has its limits. Or does it? One beetle has evolved toughness to the extreme, to the point that it can survive being run over by a car. The diabolical ironclad beetle possesses an armoured outer layer in which the wing casings (elytra) have fused together in interlocking pattern rather like dovetail joints, creating an astonishing resilience to force that would be the equivalent of us surviving having a large building dropped on us. And if this wasn't impressive enough, the entire cuticle is made of tissue layers that flex like shock absorbers and can heal themselves if they do crack. This, combined with a flattened profile, gives the diabolical ironclad beetle reinforcement qualities that are the envy of structural engineers and have scientists scratching their heads.

Io Moth

Automeris io **| Lepidoptera / Saturniidae**

The gorgeous io moth is one of North America's favourites. This is a male.

The world is replete with beautiful moths, everywhere from backyards to tropical forests. One of North America's favourites is the io moth, sharing the name of a mythical Greek princess. The male is a buttery yellow and the female a reddish-brown, but both sexes have enormous spots, almost like a peacock's eyes, that are quickly opened as a shock tactic to ward off predators.

The attractive adults are on the wing in spring in the north, and later on the caterpillars appear. These are notorious for their stinging spines, which can be painful to humans even with minimal contact. These are the most familiar stinging caterpillars in North American gardens.

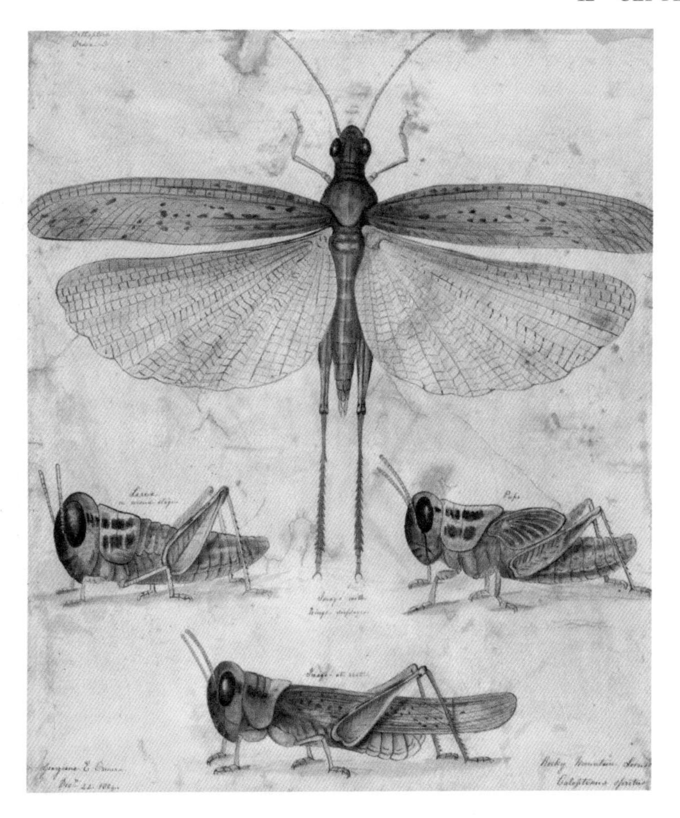

Rocky Mountain Locust

Melanoplus spretus | **Orthoptera / Acrididae**

Illustrations are all we have left of the Rocky Mountain locust.

Below are quotations from pioneering farmers from Kansas in 1874, referring to the superabundance of Rocky Mountain locusts. A swarm from 1875 (Albert's Swarm) was estimated anywhere between 3 and 12 trillion insects, perhaps the highest ever estimated gathering of animals anywhere on Earth. Forty years later, the Rocky Mountain locust was extinct.

'They looked like a great, white glistening cloud, for their wings caught the sunshine on them and made them look like a cloud of white vapor.'

'I never saw such a sight before. This morning, as we looked up toward the sun, we could see millions in the air. They looked like snowflakes.'

13th SEPTEMBER

African Dung Beetle

Scarabaeus satyrus
Coleoptera / Scarabaeidae

Today's lesson is that, however lowly you might feel, everybody can look up at the stars. The African dung beetle collects dung into balls and rolls them away from rivals to a safe distance. It needs to roll them in a straight line, for which, amazingly, it uses the night sky to orient itself. It uses polarized light from the moon and the position of the Milky Way, which appears to it as a band in the night sky. By maintaining a consistent angle to the firmament, it maintains a straight path.

14th SEPTEMBER

Hera's Gadfly

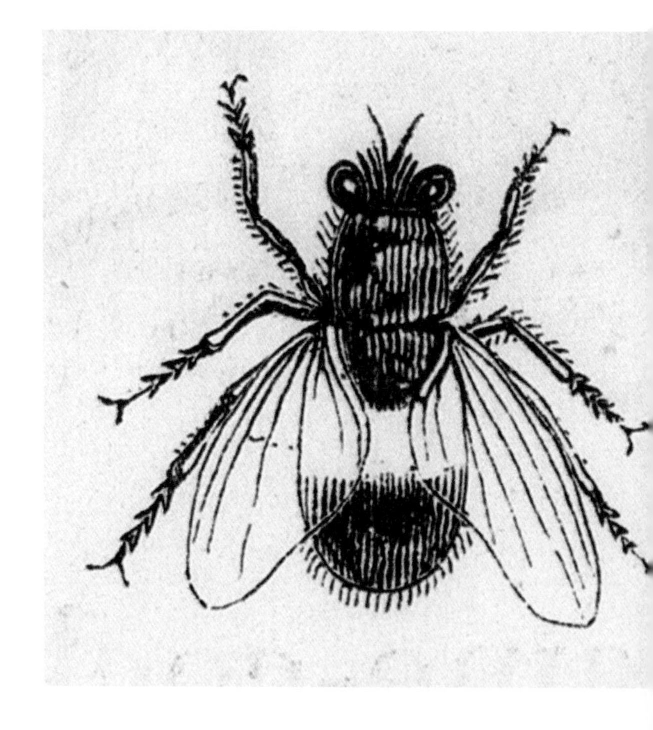

Tricked into marrying Zeus when he exploited her instinct to nurture animals, Hera also endured his persistent infidelity. After unsuccessful attempts to exact revenge on her husband, she instead turned her attention to his lovers. She turned Callista into a bear and Antigone into a stork. With Io she doubled down by transforming her into a cow, and then turned the ghost of Argus Panoptes into a gadfly and sent him to plague the newly bovine Io.

Above: Deep down (30m+) in the waters of Lake Tahoe in the USA's Sierra Nevada lives an insect that never leaves the water at any life stage.

Opposite top: The African dung beetle orientates itself using the Milky Way.

Opposite bottom: The pesky gadfly was a tool of Hera's revenge on her husband and his lovers.

Lake Tahoe Stonefly

***Capnia lacustra* | Plecoptera / Capniidae**

Stoneflies are a small and overlooked order of only 3,500 freshwater species found in almost every corner of the world. They don't have a complete metamorphosis, but have various nymphal stages, and they are generally simple bodied and primitive. They don't tend to create headlines, but one species, the Lake Tahoe stonefly from its eponymous lake that straddles the border of Nevada and California, is perhaps one of the most extraordinary of all insects.

For one thing, it lives its entire life underwater; the adult stages have no need of wings. Furthermore, it lives deep down, between 30m and 90m (100–300ft), among vegetation. And amazingly, it hatches two 'cohorts' of young a year. The May cohort are laid as eggs on waterweed. The November cohort hatch inside the female and are birthed as live young. During the summer they float freely around inside the female's body, sometimes even reaching the head!

European Hornet

***Vespa crabro* | Hymenoptera / Vespidae**

Despite having a fearsome reputation, the European hornet is a surprisingly placid and passive wasp.

The European hornet queen is a formidable looking beast, around 3cm in length. Hornets are widely feared by humans, but in reality, they don't like confrontation with animals larger than themselves, actively avoiding conflict. They prefer to go under the radar, keeping themselves to themselves with an undeviating route from nest to hunting ground – they have no time to waste causing trouble with the neighbours. The queen dedicates herself to building a nest in a rotten tree, or underground cavity, and then cycles between egg laying and hunting for provisions for her imminent arrivals. She is highly intelligent and if she senses threat, she does not attack indiscrimately but begins with communication. She lifts one of her front legs up, then flicks it several times, as if to kick a football repeatedly. Anything – or anyone – with any degree of intelligence will not stick around to find out what happens if this calm, but very meaningful signal, is ignored.

Cranefly
Diptera / Tipulidae

Craneflies look awkward, but they've been around for 245 million years.

Her jointed bamboo fuselage,
Her lobster shoulders, and her face
Like a pinhead dragon, with its tender moustache,
And the simple colourless church windows of her wings
Will come to an end, in mid-search, quite soon.
Everything about her, every perfected vestment
Is already superfluous.
The monstrous excess of her legs and curly feet
Are a problem beyond her.
The calculus of glucose and chitin inadequate
To plot her through the infinities of the stems.

From 'A Cranefly in September' by Ted Hughes (Season Songs, 1975)

Asian Swallowtail Butterfly

Papilio xuthus **| Lepidoptera / Papilionidae**

This delicately monochrome Asian swallowtail contrasts beautifully with the colourful flowers it visits.

Swallowtail butterflies are found throughout the world, distinctive with their streamer-like 'tails' and, buttery yellow wings etched with elegant, charcoal markings. The larva is a plump, emerald sausage with false eyes set back so far back on its body that it appears to have a truly enormous head. Its actual head is tucked safely underneath its pudgy folds, and here lurks another surprise: a hidden inflatable appendage called an osmeterium, which is bright orange, forked, stinky, and can be inflated rapidly like a balloon animal to confound assailants. And that's not all. In 1980, Japanese researchers used the Asian swallowtail as the basis of their research in the discovery of photoreceptors (light detecting cells) on butterfly genitalia, which help adults to hit the mark during copulation. It brings a whole new meaning to 'finding your way in the dark'.

House Fly

Musca domestica | Diptera / Muscidae

The housefly probably has the widest distribution of any insect in the world.

How if that fly had a father and mother?
How would he hang his slender gilded wings
And buzz lamenting doings in the air.
Poor harmless fly,
That with his pretty buzzing melody
Came here to make us merry, and thou hast killed him.

From *Titus Andronicus* by William Shakespeare (c. 1588)

Parent Bug

Elasmucha grisea | **Hemiptera / Acanthosomatidae**

These fine young shieldbug nymphs have benefited from the best possible start in life.

Most insects live brief, dangerous lives and quickly abandon any eggs they lay to whatever fate befalls them. One exception is a very common shieldbug from Europe, which lives in birch woods. Female parent bugs lay a diamond-shaped stash of eggs on a leaf and then stand over them to offer protection until they hatch. Should insect predators or parasites approach, the mother will flutter her wings in threat and, if necessary, release a foul-smelling fluid from her abdominal glands, which appears to be a significant deterrent.

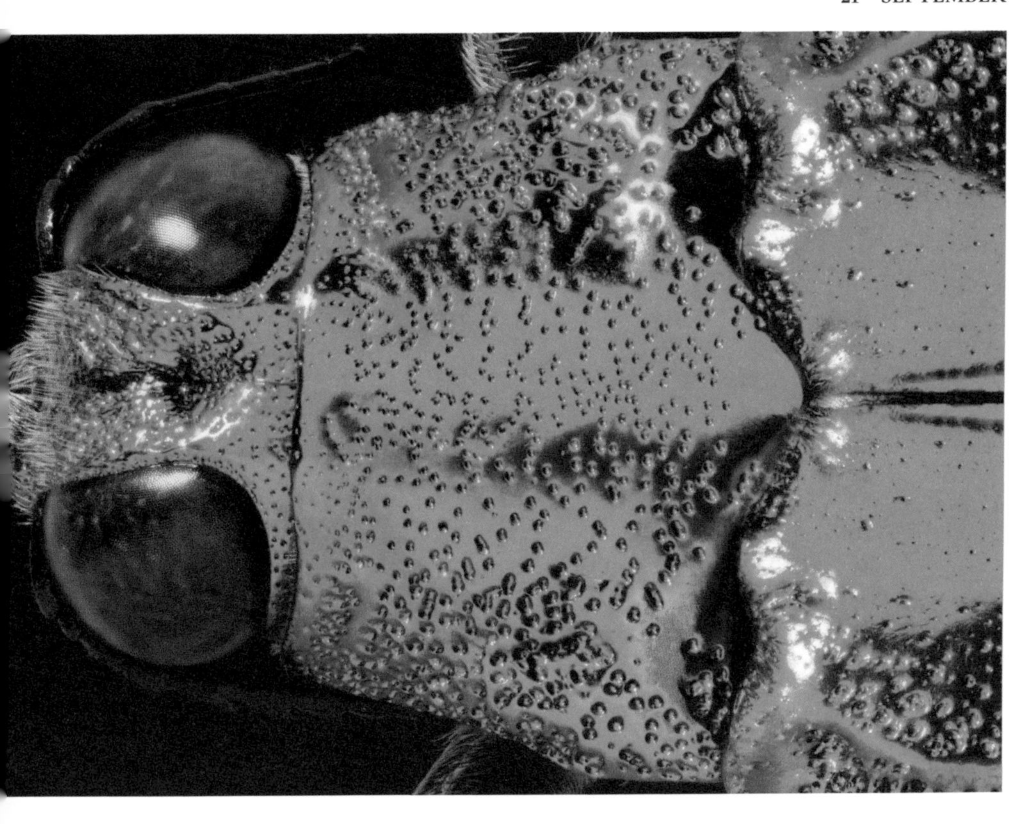

Jewel Beetle

Chrysochroa fulminans **| Coleoptera / Buprestidae**

Iridescence is a surprisingly effective form of cryptic camouflage, seen here in this close-up of a jewel beetle's head.

Camouflage: an evolutionary mechanism that has helped animals avoid being eaten for hundreds of millions of years. Whether it is the fawn-coloured antelope in long savannah grass or the mottled brown grouse on a craggy hillside, effective camouflage involves muted colours and subtle patterns that blend in with the background, right? But what about animals that are predated by animals who see differently to us? Beetles, for example, are often predated by birds, who, unlike us, can see the ultra-violet spectrum. Research is discovering that bright metallic colours are effective camouflage against avian predators by making the beetles less perceptible to them. Bold, contrasting patterns may help to break up the outline of the body, further adding to the confusion, especially for young birds who are still learning to feed themselves.

Brazilian Mound-building Termite

***Cornitermes cumulans* | Blattodea / Termitidae**

The cerrado of Brazil is dotted with termite mounds.

This termite is very common in the Brazilian *cerrado* savanna, where its conical mounds reach a density of 55 per hectare (22 per acre). On the first day of heavy rain in September, each colony releases many thousands of flying individuals (alates) to take part in a once-a-year nuptial flight. Preparations for the great day are meticulous. The workers build an exit ramp 4–12cm (1½–4¾in) long, from which the alates can launch. The exit ramps are roofed until the last minute. The great reveal ensures that the air is saturated with termites, and they cannot all be eaten by predators.

Giant Woodwasp

Urocerus gigas **| Hymenoptera / Siricidae**

An antique engraving of giant woodwasps in their various life stages.

Sawflies are a primeval groups of insects that wasps and bees are descended from. One of the largest is the Giant Woodwasp. Also known as a horntail, it can look a bit scary but this huge hornet mimic is harmless. It uses its formidable ovipositor to lay its eggs in dead or dying pine wood. The larvae then spend up to five years eating through the wood (they are known to eat small invertebrates too), before emerging as adults. A way to distinguish sawflies from wasps is their thickened 'waist' (where the thorax joins the abdomen). By contrast wasps have a tiny, constricted waist, called a petiole. Interestingly, the Giant Woodwasp has evolved a small white patch at the top of its abdomen which, at the right angle, makes it appear to have a petiole and possibly make predators think twice – clever stuff from this gentle giant.

LARGE YELLOW UNDERWING

Noctua pronuba | Lepidoptera / Noctuidae

On autumn nights, vast numbers of large yellow underwing moths migrate south.

It's September, and many nature enthusiasts will be seeing birds such as swallows (*Hirundo rustica*) gathering on wires prior to migrating south. Birds are justifiably famous for their travels, but what many people don't realize is the existence – and extent – of insect migration.

Take the large yellow underwing, an abundant late summer and autumn moth. Studies in central Europe have found that they migrate south-west in autumn in their hordes – they can even be detected on radar. The extent of the movement is unknown, but it must be vast. The moths fly at 250–500m (820–1,640ft) above ground on calm nights and can travel at more than 50km/h (31mph).

Studies have found that they use magnetic fields to orientate themselves, and probably also navigate by the rotation of the stars.

Orange-legged Dronefly

***Eristalis flavipes* | Diptera / Syrphidae**

This dronefly is just one of many bumblebee mimicking insects in the world.

Harmless hoverflies are known for being effective mimics of the better-armed wasps and bees. The orange-legged dronefly, of North America, is a fantastically fuzzy bumblebee mimic with its dense, chitinous pile arranged in broad bands of yellow and black. So effective is this pretence, that humans are easily tricked into mistaking them for bees, but look closely – there are a few key features that betray their true identity. First, bees have two sets of wings, whereas flies have only one pair while dronefly wings have a characteristic U-shaped vein on each wing, known as the Eristalini loop. The other giveaways are those big, round eyes that almost fill the head and the soft, spongy mouthparts, which suck up nectar and pollen.

CHARLES HENRY TURNER
Entomologist

A portrait of Charles Henry Turner.

Professor Charles Henry Turner was born on 3rd February 1867 in Cincinnati, Ohio. He graduated from the University of Cincinnati in 1891 with a Bachelor's degree in Biology, following up with a Master's degree in 1892 and a PhD in 1906, becoming one of the first African Americans to receive a doctorate. During his postgraduate studies, Turner began to research insects, and became prolific in his studies of the hymenoptera, particularly ants and bees. He demonstrated that honey bees can see in colour and recognise patterns, and that they retain visual memories of their environment. Following his own studies, Turner became a professor in various institutions, including Clark University, Georgia, where the Turner-Tanner Hall is now named in his honour.

Hornet Robberfly

Asilus crabroniformis **| Diptera / Asilidae**

The hornet robberfly is a supreme predator of grazing pasture.

Cow dung is not everyone's cup of tea, but there are myriad insects who welcome the sound of poo hitting pasture, and throng to it to feed, mate and reproduce. The hornet hoverfly is one of the most regular frequenters of the cowpat; a large straw-coloured fly with an abdomen that is dark on the basal half and yellow towards the tip. It has brown-tinged wings and facial hair of the kind not seen since ZZ Top last toured. That long, bristly moustache is distinctive in hoverflies, and it is frankly amazing that they can keep themselves so clean, given their propensity to hang out in wet poop. The larvae develop in or under cowpats, where they predate other coprophagous insects. Adults are one of the largest flies in the UK, reaching 2.5cm (1in) in length. Those huge, hemispherical eyes are adapted for all-round vision when hunting, so any prey in their sights will need to have eyes in the back of their own heads.

28TH SEPTEMBER

Pitcher Plant Midge

Metriocnemus knabi
Diptera / Chironomidae

The pitcher plants (*Sarracenia*) are a group of carnivorous plants with insect-trap leaves. They are trumpet-shaped and have a vertical tube with an opening at the top with a slippery surface. Insects are attracted by nectaries at the lip, they then fall into the digestive fluids that fill part of the tube. Amazingly, there is a midge that lays eggs in the pitcher and whose larvae live within the digestive fluid, competing with the plant to eat the carcasses of fallen insects!

29TH SEPTEMBER

Spongillafly

***Sisyra* spp.**
Neuroptera / Sisyridae

A spongillafly lays a cluster of eggs on a leaf, overhanging a pond, then constructs a beautiful, latticework protective structure around them. The tiny larvae are parasitic; when hatched they drop into the water and set about the task of finding their new hosts: freshwater sponges. They latch onto the outside of a sponge with specially modified mouthparts and slowly extract the innards.

MACHACA

Fulgora laternaria **| Hemiptera / Fulgoridae**

Above: Machacas have fabulous peanut-shaped heads.

Opposite top: Amazingly, an insect lives and breeds in pitcher fluid, the insects' graveyard.

Opposite bottom: Spongillafly larva feed on freshwater sponges with specially modified mouthparts.

This large plant bug is also called the peanut bug, because of its extraordinary bulbous head, which really does resemble a peanut, as you can see. The wings are equally impressive, the underwing showing huge black, yellow and white spots that look just like the eyes of an owl. All in all, it is a formidable-looking insect that induces much alarm among humankind, despite the fact that it is entirely harmless.

La Timie-perle

***Axia margarita* | Lepidoptera / Cimeliidae**

One of the world's most beautiful moths resides in southern France and Spain.

The range of tropical moths is incredible, but how about this stunner from the Mediterranean as one of the most beautiful in the world? It doesn't have a proper English name, but the French one, *timie-perle*, 'gem loved by God', will do just fine. It occurs in the garrigue vegetation of a few parts of southern Europe, and its main food is the spurge *Euphorbia duvalii*. It is fairly common, and has a long flight time, from April right through to October.

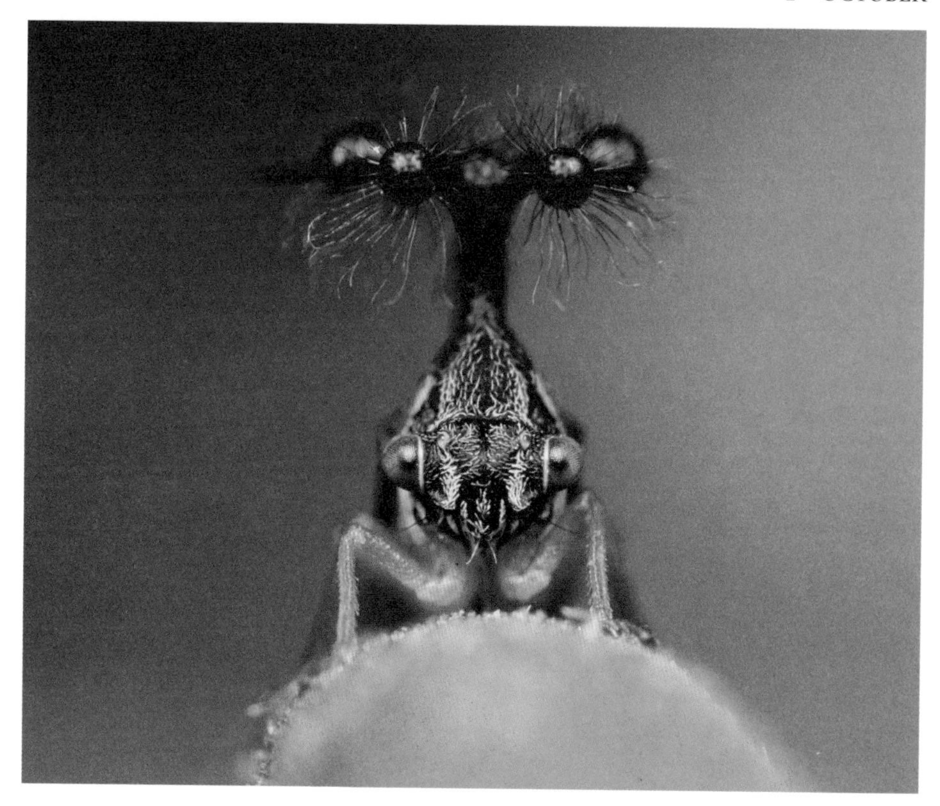

Brazilian Treehopper

Bocydium globulare **| Hemiptera / Membracidae**

Hats off to the funkiest treehopper in the Amazon.

Hats off – quite literally – to the Brazilian treehopper, which surely wins the award for nattiest headwear. A stout trunk ascends from the bug's head, branching out into four stems that terminate in large, hairy spheres. As if that isn't bonkers enough, a fifth appendage stretches out behind like a tail streamer. Its purpose is still not conclusively agreed upon, but this creative addition does resemble the fruiting body of a parasitic fungus that infects this treehopper, and so it could be an elaborate artifice to make the hopper appear inedible. Whatever it is, it is unlikely to aid camouflage – this bug is not trying to blend in. Part-interstellar spacecraft, part-fidget spinner: Philip Treacy would be proud.

Gliding Ant

Cephalotes atratus | **Hymenoptera / Formicidae**

Gliding ants have no problem with heights.

When is a flying ant not a flying ant? (see Flying Ant Day, page 224). The answer is when it's a gliding ant. This is a wingless species that lives high in the forest trees of South America, usually a long way from the ground. When their colony is threatened by a predator, the workers attack it and, not surprisingly, many are dislodged. They begin a long descent to the ground and if they reach it, they will be unable to find their way back. These ants have evolved a way of controlling their fall in mid-air, putting their abdomens up like a parachute and gliding back to the trunk further down, still well above ground. Some ants use their gliding to escape predators, deliberately falling in an emergency.

Lolonandriana (Madagascan Sunset Moth)

Chrysiridia rhipheus | Lepidoptera / Uraniidae

Lolonandriana was a taxonomic conundrum to entomologists for some years, as it was initially misidentified as a butterfly.

Native to Madagascar, Lolonandriana is also known as the Madagascan sunset moth. The Malagasay people believe the souls of their ancestors appear in the form of this dazzling moth. For some time, due to the incompetence of European collectors in the 18th century, it was classified as a butterfly (heads fell off a collection of specimens and heads were put back on the wrong bodies). Lolonandriana's actual head has long, soft and slightly feathery antennae, characteristic of moths, but in many respects, it does indeed look and behave like a tropical butterfly. It is day-flying, has forward facing wings that form a 'V' shape, and long streamers on the hind wings. The intense colours are aposematic: toxins within the larval foodplant *Omphalea* are metabolized by the caterpillars, rendering themselves and the resulting adults inedible – a warning to predators is advertised by the intensely iridescent wings, which are also a handy camouflage tool.

STABLE FLY

***Stomoxys calcitrans* | Diptera / Muscidae**

It's not just the mouthparts that make stable flies a force to be reckoned with – it's their tenacity.

The ancient Egyptians were famous for their worship of scarab beetles (see page 48), but what is less well known is that they also had a sneaking admiration for flies. The stable fly was a constant irritant to the people of the New Kingdom of Egypt (1550–1069BCE), continually biting their cattle and annoying people too. But however much they tried to persecute flies and get rid of them, the recalcitrant *calcitrans* just kept on coming back for more. Bravery and bloody-mindedness are favourable military attributes, hence the Order of the Golden Fly that was created to commemorate acts of valour. Some original golden fly jewellery survives today.

Rainbow Jewel Beetle

Chrysochroa fulgens **| Coleoptera / Buprestidae**

Nearly every colour of the spectrum is visible on this extraordinary rainbow-coloured beetle.

Here's another jewel beetle species to knock your socks off. *Chrysochroa fulgens* is a beetle of astounding artistry. All along its length, deep crimson spots radiate outward and bleed into yellow, then green and finally blue. It is a walking, flying chromatography experiment. A thick band of cream, like burnished bone, traverses the elytra creating a startling contrast to the lustrous rainbow. It is native to Thailand and, like other jewel beetles, spends the majority of its life hidden from sight as a small, wood-boring larva, snacking its way through the sapwood of trees before pupating, whereupon it makes its magnificent debut into the outside world.

Violet Dropwing

Trithemis annulata **| Odonata / Libellulidae**

This dragonfly drops its wings to rest.

It's the subtleties of this dragonfly that make it attractive. The violet colouration on the male's abdomen isn't excessively garish, it's just right, and it goes perfectly with the orange wash to the wings and the amber wing bases. If it were a sports car, the livery would be the most expensive part.

The name 'dropwing' describes this insect's very distinctive habit of drooping the wings at rest while lifting up the abdomen into the so-called 'obelisk' position. It is thought that the dragonfly's abdomen is deliberately pointed to the sun to minimize the area of exposure to the sun's rays.

The violet dropwing is a very abundant tropical African dragonfly, which has been spreading northwards in recent years. It was first recorded in Europe in the late 1970s (Spain), France in the 1980s and mainland Italy in the 1990s.

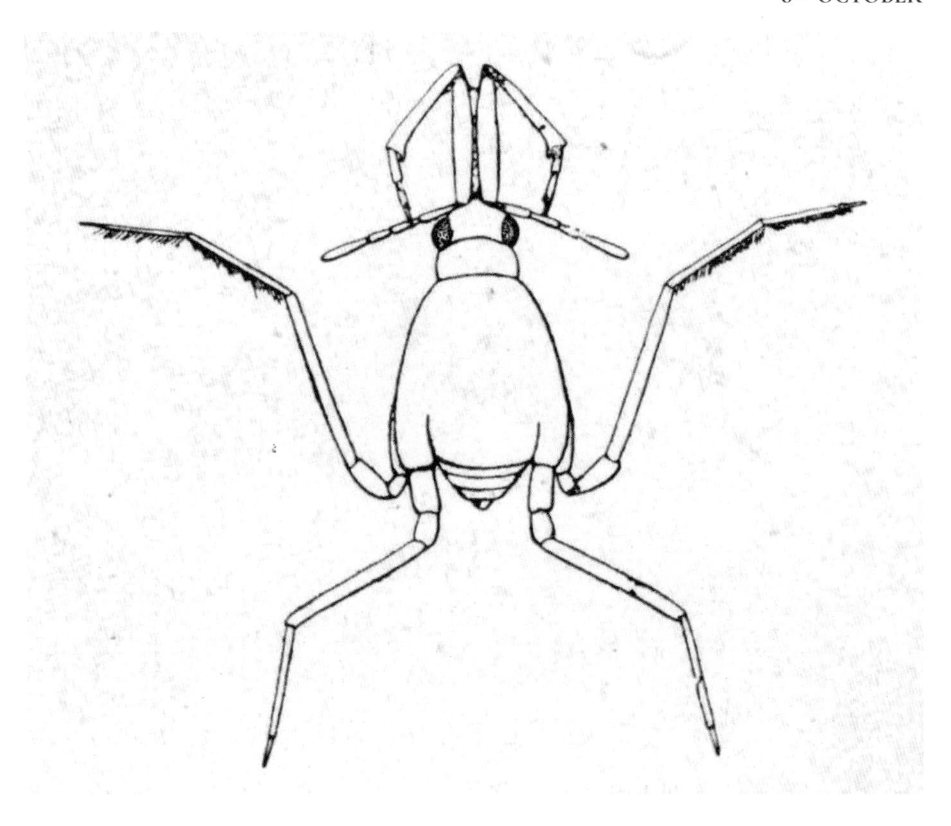

Sea Striders

***Halobates* spp. | Hemiptera / Gerridae**

The sea striders are one of the few insects that can live on open sea.

On a planet where insects dominate the air and land, it is surprising that only one tiny group has conquered the sea. *Halobates*, also known as the sea striders, is a genus of water skater that has shirked a terrestrial existence, choosing instead to live a life on the waves. There are numerous species of sea striders that dwell in and on tropical and sub-tropical coastal waters, and five known species live out on the open ocean. Their long legs sit on the surface tension, enabling them to skate across the water without sinking. Sea striders eat zooplankton and reproduce by attaching egg clusters to passing flotsam and marine debris.

MILLIPEDE ASSASSIN BUG

***Ectrichodia crux* | Hemiptera / Reduviidae**

Millepedes cannot make easy prey for anything, but they have vulnerable spots.

There are few more evocative names than 'assassin bug', and these live up to their name, being among the very few invertebrates that specialize on millipedes, which are armed with powerful chemical defences and tough exoskeletons. Nevertheless, the bugs ambush them, approaching furtively at first and then rapidly grabbing the intended meal with their claws and inserting their needle-like rostrum into a spot near the rear. In contrast to many bugs, which simply suck the fluids at this point, these assassin bugs inject powerful paralytic toxins in their saliva. They also inject fluids to begin the process of digestion.

These hunters are active at night and, if one immobilizes a large millipede, others, including the bright red nymphs, come and join the feast.

Termite-mimicking Beetle

Austrospirachtha carrijoi | **Coleoptera / Staphylinidae**

This beetle has taken mimicry to extreme lengths.

Mimicry typically involves the modification of one or more parts of the anatomy to resemble another animal, such as a constricted waist, or bold stripes. Such methods for passing oneself off as something else take thousands, sometimes millions, of years to evolve, so imagine the biophysics involved in growing an entire termite on your back. *Austrospirachtha carrijoi* is a rove beetle that appears to have rewritten the handbook on camouflage, going one further than mere appendage or colour. It can enlarge its abdomen into what looks like a lifesize termite-shaped balloon, substantially bigger than the rest of istelf, as though it is wearing a gigantic backpack. This has happened because a long time ago this beetle took a shine to the contents of termite nests and, as we know, disguise is key to safe passage. To avoid being attacked, it is likely that this beetle's descendants gradually adapted increasingly large, termitey abdomens, resulting in the fancy-dress-competition-winning little beast we see today.

Death's-head Hawk-moth

Acherontia atropos **| Lepidoptera / Sphingidae**

The thorax of this species does have a remarkably skull-like pattern.

One of the world's most charismatic insects, the death's-head hawk-moth has long stirred the imagination of humankind because of the curious pale marking on its thorax, which bears a striking resemblance to a human skull. It has long been associated with death and worse. The genus name *Acherontia* comes from Acheron, the River of Woe or Hell in Greek mythology, a river across which the dead are ferried. It acquired later fame by adorning the poster for the movie *The Silence of the Lambs*, a tale in which the serial killer places the larvae of the hawk-moth in the mouth of each of his victims.

The real-life moth is entirely harmless to people. The skull-like pattern vaguely resembles a bee and could be part of the moth's defence when it raids beehives for honey, one of its unusual habits. It also cloaks itself with the chemical imprint of a bee, meaning that the colony members don't immediately evict it when it is on a raid.

Predatory Bush-cricket

Saga pedo | **Orthoptera / Tettigoniidae**

The spiny, spindly predatory bush-cricket blends in expertly among *Eryngium* plants, waiting to ambush its prey.

The predatory bush-crickets are unusual in several ways. One is that they are carnivorous; whereas most bush-crickets are vegetarian, these are hunters, preying on smaller insects and sometimes even each other. Not only that – they are some of the largest insects in Europe, with adult females reaching lengths of around 15cm (6in). Much of the female's size is made up of the intimidating ovipositor, which is used to drill into the soil to lay eggs but is otherwise harmless. *Saga pedo* is sometimes referred to as the 'spiked magician', on account of the formidable spines on its legs, which help it grip onto its prey. But the most impressive trick in the predatory bush-cricket repertoire is in its reproductive lifestyle. Males are virtually never recorded; almost all records are female. Adult females reproduce asexually (without the need for male sperm to fertilize their eggs), subsequently almost all eggs laid will hatch into females.

Cactus Longhorn Beetle

Moneilema gigas | **Coleoptera / Cerambycidae**

The larvae of this smart, black beetle live inside cacti, protected from the hot, dry desert climate.

The cacti of the Sonoran Desert, which straddles the border of Mexico and the southern USA, are a hardy bunch. They grow in the hottest, driest of conditions in an environment in which little else can survive, and their resilience has opened up niches for other species to thrive. One of these is the cactus longhorn beetle, a medium-sized black beetle distinguished by lightly mottled elytra and a white band on its antennae. In its larval stage it lives within the fleshy tissue and roots of a cactus, eating away at the innards until it builds up enough fat reserves to pupate, whereupon it will emerge through the outer wall as a glossy adult to find a mate. They are not very mobile, however; cactus longhorns are flightless. Having dispensed with wings at some point in their evolution, their elytra are now fused together, and so the adults trundle around on foot hoping to find love in the desert.

Desert Locust

Schistocerca gregaria **| Orthoptera / Acrididae**

The desert locust is the only insect known to have flown across the Atlantic east to west.

Desert locusts are known for their swarms, or 'plagues'. Every so often, in response to environmental conditions, they swap from a solitary form to a highly gregarious and movable form. Then they can gather in vast numbers, sometimes 150 million per sq. km (1/3 sq. mile), and can travel at the speed of the wind, sometimes being wafted high into the air.

These feats are mightily impressive. In October 1988, a swarm was blown off the coast of West Africa and right across the Atlantic to the Caribbean, a distance of 5,000km (3,100 miles). Somehow the insects survived the six-day journey, and nobody is quite sure how they did it!

iCikwa

Sphaerocoris annulus **| Hemiptera / Scutelleridae**

Above: The iCikwa makes a bold statement.

Opposite top: Even some keen insect enthusiasts have never heard of webspinners.

Opposite bottom: The call of this cicada is neatly imitated in the local name nyenze.

The iCikwa is a bug native to the tropics of Africa. It is a shieldbug, or stinkbug, on account of its ability to emit a foul-smelling chemical when threatened or attacked. It is also known, in the west, as the Picasso bug, because of the incredible markings on its exoskeleton. The abstract brushstrokes of colour (here, turquoise and red), outlined in black, are seen in the artwork of many African cultures. Pablo Picasso became obsessed with this brave, vibrant approach to artistic expression, using it in his own work to rail against the realistic conformity and convention of European art in the early 20th century. This style eventually became known as Cubism and planted itself firmly in the modern art movement.

WEBSPINNERS
Embioptera

Hands up who's heard of a webspinner? And we don't mean a spider, either. The webspinners are an obscure order of insects that, you've guessed it, spin webs, which they do from a gland on the forelegs. The thick webs aren't draped across spaces, but are spun prone to the ground or across a tree trunk to make 'galleries' that provide protection from predators. Females dote over their eggs and nymphs, some covering each egg with mashed-up plant material to camouflage it.

17TH OCTOBER

ORANGE-WING CICADA
Platypleura haglundi
Hemiptera / Cicadidae

Anybody who is familiar with moths will immediately be struck by the similarity between the hindwings of the orange-wing cicada and the underwing moths. A threatened animal can suddenly show the bright colours and give an enemy enough of a shock for it to make a getaway. This species is common in the southern African bush, and the local Shona name *nyenze* gives an idea of the song.

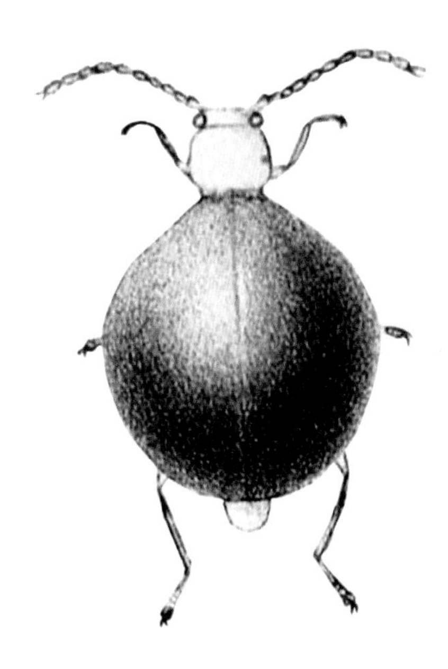

Choresine Beetles

Choresine spp. | Coleoptera / Melyridae

You may have never heard of this beetle, but the frog that eats it is infamous.

The rainforests of South America are home to a beetle you've probably never heard of, but it has a very famous associate that you most certainly will have. *Choresine* species are soft-winged flower beetles that look perfectly innocuous, but hide a deadly secret, for they contain batrachotoxin, a potent neurotoxin that, even in the tiniest doses, is lethal to humans. These beetles make up a significant part of the diet of *Phyllobates terribilis*, aka the golden poison frog. This tiny, bright yellow frog, endemic to the rainforests of Colombia, is famous for being a source of the poison that Indigenous people use to coat their hunting spears. Research has discovered that, under laboratory conditions, the frogs contain no toxins, and so it is believed that the toxins they contain in the wild are a direct result of bioaccumulation of toxins from the *Choresine* beetles they ingest. This in itself is remarkable, as this tiny frog has evolved to metabolize quantities of batrachotoxins that would floor an army.

Merveille Du Jour

***Griposia aprilina* | Lepidoptera / Noctuidae**

This lovely moth is a favourite autumn special for moth enthusiasts in Europe.

October represents the dying of the light of the moth year, when in temperate areas the variety of active species falls sharply. There are fewer balmy nights to enjoy, to put out the non-lethal light trap and see what the lepidopteran post will bring. But there are compensations. Many of autumn's species are gorgeous, resembling colourful leaves, clad in yellow, orange or red. Their names are evocative, too: how about frosted orange, pink-barred sallow and beaded chestnut? But no moth brings autumn cheer quite like the glamorous merveille du jour. This is a predominantly green moth, of a fresh but subtle hue, with a latticework of black streaks, blotches and speckles that fits perfectly against a background of moss or lichen. It might be an autumn moth, but the green is almost springlike. The evocative French name means 'the best thing I've seen all day'. How perfect.

Sawfly

***Pseudoperga lewisii* | Pergidae**

This sawfly diligently guards her young to increase their chances of survival into adulthood.

Beyond social bees and ants, we don't tend to think of insects as being active parents, but there are a surprisingly large number of species that take their parental duties very seriously. *Pseudoperga lewisii* is a small, pale sawfly from eastern Australia and Tasmania, which stays with her brood long after she has laid her eggs. She stands over the newly hatched larvae as they eat, protecting them from potential predators and parasitoids. This behaviour is known in several sawfly genera, which use a variety of methods to look after their young, including aposematic colouration and defensive postures.

Giant wētā

***Deinacrida fallai* | Orthoptera / Anostostomatidae**

The wētā is endemic New Zealand, and the heaviest insect in the world.

Giant wētā, or wētāpunga, are endemic to New Zealand. These enigmatic orthopterans are one of the largest insects on Earth, reaching around 70mm (2¾in) in length and weighing as much as a mouse. In order to reach these generous proportions, they go through an astonishing 11 stages of moulting, shedding an outer layer with every growth stage. Wētāpunga are nocturnal – they feed and breed by night, roosting in leaf litter during the day. Human-driven habitat loss, and predation by invasive species introduced by European colonists, has wiped the giant wētā out on the mainland. The small populations that have survived on a few outlying islands are slowly recovering, supported by carefully monitored captive breeding and translocation programmes.

Common Water Strider

Aquarius remigis **| Hemiptera / Gerridae**

The world of the water strider rests on surface tension…

The water striders, or pond skaters, are a group of specialized bugs that live on the interface of air and water, using their long legs to ride the surface tension. They are carnivorous, attracted to the waves made by insects that have fallen into the water and are struggling. The predator quickly grabs the prey and injects its rostrum, sucking out the juices.

Water striders also communicate using surface tension. They have a code of waves. If a male produces surface ripples at 80–90Hz, this is a warning to a rival to back off. If it ripples at 3–10Hz, that's an invitation to a female.

American Cockroach

Periplaneta americana | **Hemiptera / Blattidae**

It is said that cockroaches will survive the apocalypse, but will we survive with them?

No insect book is complete without a nod to what is arguably the most notorious species of all – the American cockroach. It is a bug that has wound its way into our homes and psyches. It sets us on edge and is to many the very definition of 'creepy-crawly'. However, if we look at this insect from a more pragmatic perspective, we see an extraordinarily adaptable animal that rivals (and possibly surpasses) humans in its reproductive and survival success globally. The American cockroach is native to Africa and the Middle East, but was introduced into the Americas by way of Transatlantic slave and trade ships in the 17th century. Since then, it has proliferated, thanks to its ability to eat almost anything and survive most conditions.

JAPANESE DRAGONFLY ART
Odonata

The Dragonfly and the Bellflower, by Katsushika Hokusai, late 1820s.

What do you associate with Japan? Its amazing cuisine perhaps? Japanese gardens and blossom? The Samurai? It's unlikely that dragonflies are high on your list, but these insects are deeply revered on the islands. Indeed, an alternative name for Japan itself is Akitsushima, 'Island of Dragonflies'.

Part of this is down to the rice industry, because dragonflies are drawn in huge numbers to paddy fields. In the past, they were thought of as the rice spirit and a good luck charm for harvest. The general good luck theme is echoed in their art.

The Samurai must have been keen observers, because they associated dragonflies with speed and decisiveness, which is certainly apt. Their warriors frequently adopted dragonfly emblems on their battle armour.

Malaysian Exploding Ant

Colobopsis saundersi | Hymenoptera / Formicidae

These ants literally do explode, their body fracturing when threatened.

Ants are famous for their commitment to colony life. At times of danger, they can be notably self-sacrificial in defence. The most remarkable example occurs in the Malaysian exploding ant, which does exactly what its name implies.

The workers have huge jaws that are connected to a relatively enormous mandibular gland. This is filled with sticky, corrosive fluid. When the ants are in peril from predators, or the colony is threatened by rival ants, the situation can become desperate. Left with no alternative, the worker can tense its muscles and rupture its abdomen, ripping its limbs apart and spraying the gland fluid everywhere. Predators are startled and put off. Rivals may be fatally covered and immobilized.

Purple Pleasing Fungus Beetle

***Gibbifer californicus* | Coleoptera / Erotylidae**

Purple as iridescence is quite common in the insect world, but purple pigment is very rare, and rather pleasing.

Purple is a rare colour in the animal world, seen mainly in aquatic species but sadly all too lacking in the terrestrial domain. Apart from butterflies, very few insects display purple colouration, and in most cases the purple is the result of iridescence, in which layers of chitin scatter light randomly, reflecting different hues of colour back into our eyes. This beetle's colour, however, is even scarcer because it is actually derived from pigment, which is natural colouring embedded in its cuticle. *Gibbifer californicus* is one of several pleasing fungus beetle species that come in various hues from powder to deep blue, but by far the most pleasing (pardon the pun) is this bright violet version.

Circadian Rhythms

Atta cephalotes **| Hymenoptera / Formicidae**

Leafcutter ants form some of the largest colonies of any animals, often with 500,000+ individuals.

This week in some parts of the world, we turn the clocks forward or back on daylight saving time. Leafcutter ants can switch their activity, too.

The leafcutters live in enormous colonies with an average of 500,000 individuals. To us, that seems like a lot of ants, but to a parasite, that represents the richest of opportunity. The ants have a wealth of parasites, some of which are parasitoids, killing the hosts. Leafcutter workers usually come out by day, and so do the parasites, so when things get particularly bad, and there is a high incidence of infection, there is nothing else for it. The colony gets a message, and the ants' main activity switches so that they send out foragers mainly at night-time.

SAINT HELENA GIANT EARWIG

Labidura herculeana

Dermaptera / Labiduridae

The world's largest earwig once occurred on the island of Saint Helena, which is a dot in the Atlantic Ocean 1,950km (1,200 miles) west of the African mainland and 4,000km (almost 2,500 miles) east of South America. Somehow once an earwig reached there and evolved into a giant, which isn't unusual for island forms that have no predators.

In 1502, Portuguese invaders arrived and destroyed the lush habitat, also introducing invasive plants and animals. The giant earwig, discovered in 1798, was probably eaten by introduced rodents, and the boulders under which it lived were requisitioned for building work. It was last seen in 1967 and was declared extinct in 2014, belatedly being described as 'the dodo of Saint Helena'.

Above: The eyed ladybird has striking pale haloes around its black spots.

Opposite: The world's largest earwig, 8cm long, is now sadly extinct, last seen in 1967.

EYED LADYBIRD

Anatis ocellata **| Coleoptera / Coccinellidae**

Reaching almost 1cm (½in) in length, the eyed ladybird is the largest ladybird in Europe. It is a handsome beast, resembling a robust seven-spot with its bright red wing casings, but with pale haloes around the spots and white markings on the pronotum (although one or both of these features is occasionally absent). The eyed ladybird can be found on coniferous trees throughout the Palearctic region, where both larvae and adults are enthusiastic consumers of pine aphids.

Trap-jaw Ant

Odontomachus bauri **| Hymenoptera / Formicidae**

These ants' jaws snap shut in 130 millionths of a second.

This is one of those insects that makes you ask: 'Is this for real?' But yes, there really is an ant that can open its jaws 180 degrees and then slam them shut with such force and speed that prey is killed instantaneously. The jaws can close at a force 300 times greater than the weight of the ant. They shut at a speed of 230km/h (143mph), and they snap together in 130 millionths of a second!

They don't just use their jaws to catch prey, but can also use them to jump in the air as an escape response. They just aim their jaws at the ground, snap them together, and they are propelled upwards – up to 40cm (almost 16in)!

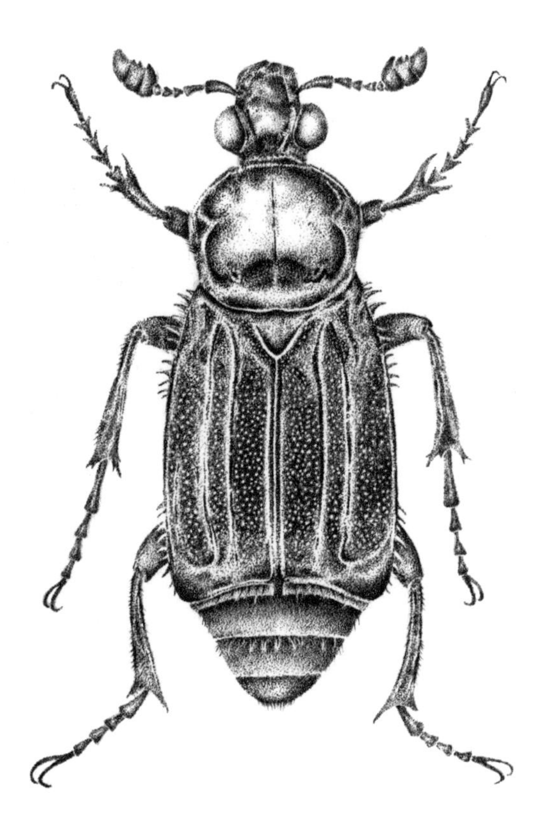

Undertaker Beetle

Nicrophorus humator **| Coleoptera / Silphinae**

The undertaker beetle plays an essential role in the removal of larger dead animals from the ecosystem.

What happens to dead animals in the wild? The answer is that an entire community of necrophagous organisms show up and set about consuming it. One of the largest is the undertaker beetle, so named because it is a uniform, silky black (save for orange tips to the antennae). It is one of the carrion beetles, and it doesn't take long to arrive on the scene. As soon as the flesh starts to give off a smell the undertaker beetle detects it from some distance away, even on a gentle breeze. Those funky, comb-like antennae catch a whiff and it's off in search of (not so) fresh meat. It lays its eggs in the carcass and the emerging larvae eat their fill. The cumulative effect of thousands of creatures descending on an all-you-can-eat buffet is the fairly swift disappearance of all most of the expired animal, save for the bones and hair. It may sound a bit grim and grisly, but without these guys to clean up, our planet would be a whole lot smellier.

1ST NOVEMBER

Brown-hooded Cockroach

Cryptocercus punctulatus

Blattodea / Cryptocercidae

Let's hear it for the lowly insects that people tend to despise. Let's hear it for the brown-hooded cockroach. This critter barely ever leaves dead logs, yet it is absolutely exceptional in one regard: its parenting. Male and female roach look after their young from the moment the eggs are laid until the progeny are past third instar stage. They keep the nest sanitized and protect and feed their young for more than three years.

2ND NOVEMBER

Fossil Katydids

Orthoptera / Haglidae, Prophalagopsidae

The fossil record shows that katydids were the first insects to evolve ears and a sound-producing system. Specimens from the Jurassic period about 160 million years ago show they could produce songs with frequencies ranging between 4kHz and 16kHz. They overlapped with all the best dinosaurs. What did that world sound like? What did they experience? Oh, for the ears of a katydid!

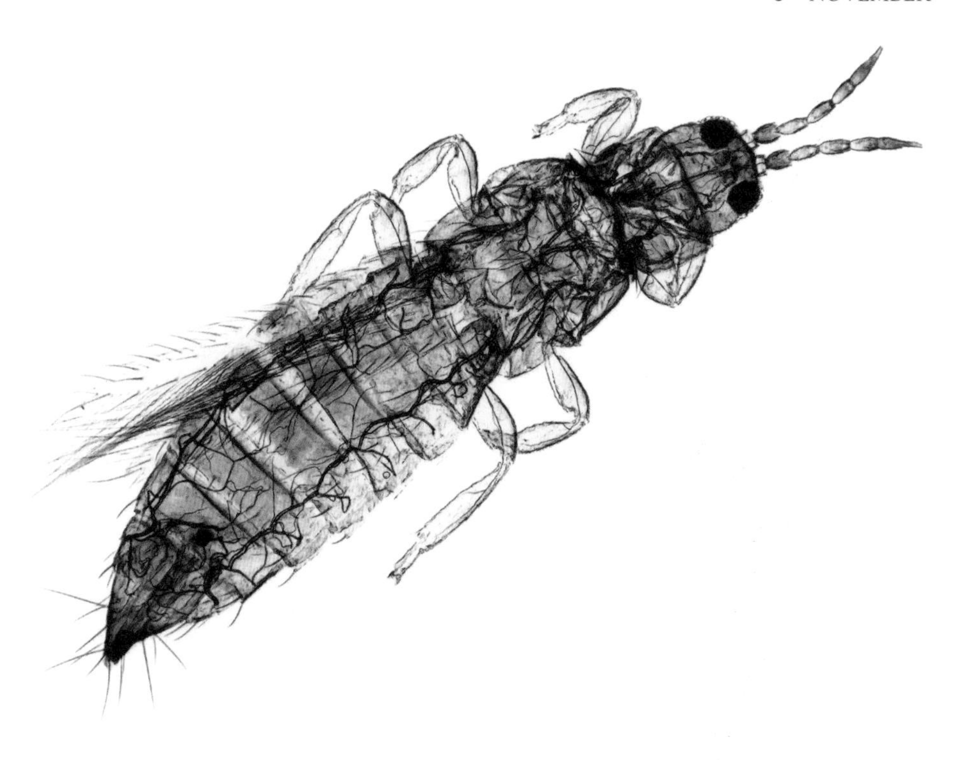

Thrips
Thysanoptera

Above: Thrips are rarely noticed and little is known about them, but they are everywhere.

Opposite top: Incredibly, male and female brown-hooded cockroaches may stay with their young for 3 years.

Opposite bottom: Some katydids were able to hear the sounds of giant dinosaurs.

There's an insect in the garden that you may not be familiar with, or you may have assumed is something entirely different, and this would be perfectly acceptable, for these guys are around 2mm (1/12in) in length and lacking any discernible features to average human eyesight. Thrips are a fairly novel insect in their own taxonomic grouping. They aren't quite a fly, not quite a beetle, nor a bug, though they could be mistaken for all of them. They are also, amusingly, called thrips in both singular and plural forms. They are Thysanoptera, a separate group that has uniquely asymmetrical mouthparts – one side is larger than the other, presumably adapted to their feeding method. Otherwise, they are fairly regular insects; they are long and slender, usually winged (although they don't so much fly, as clap their feathery wings together and leap) and like to hang out in flowers and vegetation. Thrips are largely viewed negatively due to their fondness for eating plants, including those we value, but this massively devalues this insect's place in the wider ecosystem, and from a thrips' perspective, everything's got to eat...

Red Dwarf Honeybee

Apis florea **| Hymenoptera / Apidae**

When threatened, these honey bees will work together to create a confusing racket in the face of predators.

Unlike its largely domesticated relative, the western honeybee, the red dwarf honeybee lives a resolutely wild and independent life in the forests of southern Asia. Typically a colony will construct a nest comprising a single exposed comb on a tree branch. Once the nest is outgrown, a swarm will strike out together on a reconnaissance mission to find a location for their new colony, even taking with them beeswax from the old nest to recycle into the new construction. When threatened, the first members of the colony to notice will emit a warning signal, known as 'piping'. In response, their neighbours will begin making a hissing noise with their wings, which ripples out through the colony and becomes a synchronized, cacophonous roar. The aim of this performance being to utterly discombobulate any predators that might be feeling brave (or foolhardy) enough to attempt a raid.

Spoonwing Lacewing

Nemoptera sinuata | Neuroptera / Nemopteridae

The spoonwing lacewings rival showy butterflies in colour and shape.

The spoonwing lacewings rival the world's most showy butterflies for their extraordinary wing ornaments, which look like long tails. They have a fluttering flight and look weak and defenceless, spending a great deal of time sitting on rocks or plants. The tail streamers, though, are thought to make the animals look much bigger to potential predators, and also break up the shape. The adults use their long rostrum to eat pollen.

Flesh Flies

Sarcophaga spp. | Diptera / Sarcophagidae

Flesh flies are able to give birth to live young, completely cutting out the vulnerable egg stage in their life cycle.

Flesh flies are a familiar sight in summer, and pretty unmistakeable. These large, flying humbugs have smart black and grey stripes on the thorax and a chequerboard pattern on the abdomen. They often rest on the ground in the sun where they soak up rays and clean themselves, showing those burnt sienna eyes that contrast starkly with the monochrome. Some of them also have large, flappy, paw-like feet, which can make them look rather endearing. They are also unusual in that most flesh flies are viviparous – rather than laying eggs, females birth live larvae, which means that they can get to work eating immediately rather than waiting to hatch. Despite their rather specific name, flesh flies do not eat our flesh; should they land on us they will want nothing more than to lap up salt and nutrients from the surface of the skin. They are unfussy though, seeking nutrition for everything from flowers to faeces, so it's probably best not to share your food with them.

DRIVER ANT

Dorylus wilverthi | Hymenoptera / Formicidae

Columns of driver ants are protected, as in all army ants, by the fearsome soldier caste.

Driver ants are army ants, species in which colony members form swarming raids each day, during which they can denude an area of its small invertebrates, which are killed, carted off to temporary bivouacs and fed to the growing larvae. Driver ants are mostly found in Africa, although some occur in tropical Asia, too (they are the equivalent of *Eciton* in the Americas). These colonies have a single queen that cannot fly and whose main function is to lay millions of eggs.

Dorylus wilverthi holds two ant records. It forms the world's largest colonies, with as many as 20 million workers at a time. And the queen is the largest ant in the world, acquiring a mighty length of up to 5.2cm (just over 2in)!

Madagascar Hissing Cockroach

***Gromphadorhina portentosa* | Hemiptera / Blaberidae**

The Madagascar hissing cockroach seems an unlikely candidate for the pet trade.

It's a remarkable fact that a generally derided insect (a cockroach) from the forest floor leaf litter on the island of Madagascar should become world famous. But it's true. The hissing cockroach is kept as a pet around the world and has even featured in movies. The reason is down to one thing: it hisses. And not many insects hiss.

One of the world's largest cockroaches, it makes its unusual sound by expelling air through specialized spiracles, the main respiratory organs of land insects. It's a bit like the deflating of a tyre puncture. It is used primarily as a form of self-defence against predators.

However, male cockroaches also have two other types of hiss. One is to put other males in their place. The other is an invitation to females.

Hitchhiking Wasp Larvae

Hymenoptera / Eucharitidae

The larvae of hitchhiking wasps are phoretic – they use ants to move between food sources.

The tiny Eucharitid wasp is a parasite of ants that live underground – not easy to reach, but this is a wasp with a plan. Eggs are laid into the leaf tissue of a plant frequented by the host ant, then the larvae (planidia) hatch out and hop on to passing worker ants and hitch a ride back to the nest. Eucharitid planidia are particularly motile and have a harder exoskeleton than most larvae, allowing them to pull off this hitchhiking stunt, known as phoresy. Once inside the nest, they drop off the workers and latch onto the ant larvae. Here, things get a bit dark, as the planidia eat away at the ant larvae, very carefully, avoiding critical organs and tissue; this keeps the host alive and fresh longer. Amazingly, the ant larvae survive being partially eaten, and is even able to pupate. The wasp itself pupates, and either exits the nest on its own, or leaves in the manner it came, just hitching a ride.

African Assassin Bugs

Rhinocoris spp. **| Hemiptera / Reduviidae**

Assassin bug males guard their eggs, but not for the reason you would expect.

Paternal parenting duties are not unusual among insects, however, the males in a group of assassin bugs – the Harpactorinae – have exploited childcare services for their own gain. The male mates with the female then sticks around until she lays a clutch of eggs. He then stands over them to guard them. While this does protect the clutch from predators and parasites, his main aim is to ensure that *his* sperm rather than the sperm of another male will fertilise the eggs. He continues to guard the developing eggs because, as research has shown, females are attracted to 'brooding' males. He mates with consecutive females, building quite the little stockpile of progeny around him, all in the name of increasing his gene pool.

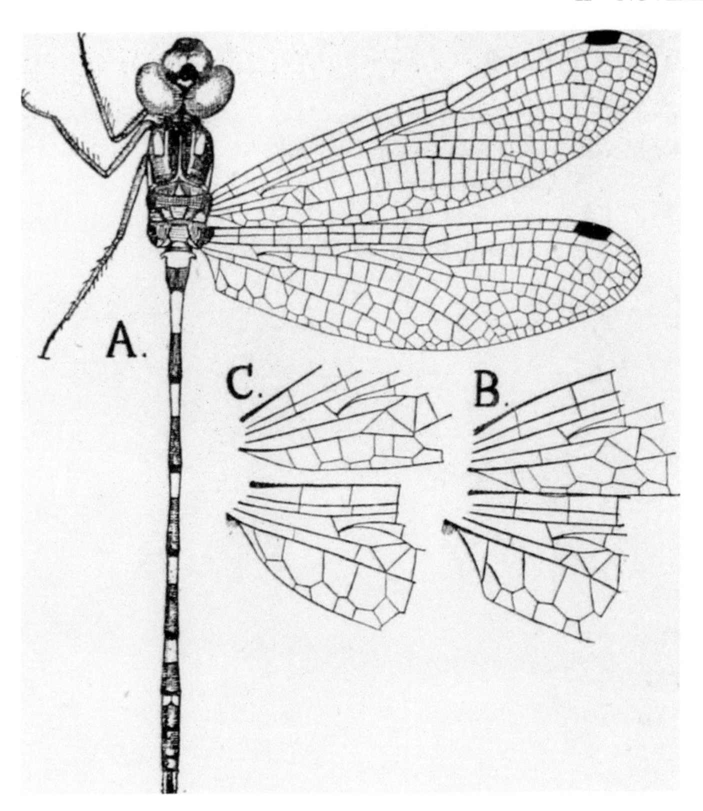

Common Shutwing Dragonfly

***Cordulephya pygmaea* | Odonata / Cordulephyidae**

Though very small, and with damselfly-style wings, the large, joined eyes of this species give it away as being a dragonfly.

Dragonflies and damselflies are typically distinguished by the way in which they hold their wings at rest. Dragons hold them out horizontally out to the side while damsel wings are held upright behind them, along the length of their bodies. There is, however, an endemic genus of dragonflies in eastern Australia that has thrown the rule book out of the window and adopted the damselfly wing method of resting its wings, giving it its name. The common shutwing is an impossibly small and delicate dragonfly, barely 2cm (¾in) in length. Indeed, the only giveaway that this is a dragonfly are those distinctive eyes – large and bulbous, which wrap right around the head and meet at the top.

WATER SCORPION

Nepa cinerea | Hemiptera / Nepidae

It isn't hard to see where the water scorpion gets its name.

Among the many impressive insects found underwater is the water scorpion, a creature with attitude that uses its large front legs for grabbing and holding prey, which includes fish and tadpoles. Somewhat incidentally, it can also deliver a painful bite to curious pond-dipping humans. Overall, it bears a striking resemblance to a scorpion, not only for the pincers, but also for its long tail, which acts as a breathing tube. It is fitted with hydrostatic receptors, which help it to maintain correct depth to breathe.

For an animal that spends most of its life underwater, the water scorpion does have an embarrassing deficiency. It is a hopeless swimmer, relying mainly on crawling about.

Plant Bug

***Heterotoma planicornis* | Hemiptera / Miridae**

This bug is a marvellous mash-up of colour and proportion.

Welcome to the weird and wonderful world of the true bugs. The plant bugs are a fairly uniform group, so it's always lovely to see a fairly bonkers change to the body plan. *Heterotoma planicornis* starts out as a slim, black bug, but then breaks convention with stocky, bright green legs. The *pièce de resistance*, however, are the antennae that were clearly made for a bug at least twice the size. This fabulous photofit beast is a plant bug, however the nymphs are partly predatory, and the adults are possibly carnivorous and even cannibalistic. It's just another example of the glorious nonconformity of the insect world.

Corn Root Aphid

***Protaphis middletonii* | Hemiptera / Aphididae**

It's common for ants and aphids to have a co-dependent relationship.

The corn root aphid and the cornfield ant (*Lasius americanus*) have a co-dependent relationship, in which the ants care for the aphids and the aphids provide honeydew for them. Honeydew is essentially aphid excreta – the bugs drink the sap from plants and exude the excess, which is very sweet. Ants get a sugar rush from it, so much so that they farm its producers like cattle. In the autumn, ants gather aphid eggs and store them underground to survive the winter. In the spring, they personally carry the aphids first to an early-growing host called smartweed (*Persicaria amphibia*), and then to corn roots later on. Here they milk them and protect them from all predators.

Montane Iris Skipper

***Mesodina aeluropis* | Lepidoptera / Hesperiidae**

You can find this butterfly in the Blue Mountains of Australia.

One of the many things that us insect lovers adore about our six-legged heroes is how beguilingly picky they can be. There are unnumbered species around the world that only occur on one plant, or in one location, at a certain altitude.

Here's a delightful example. The montane iris skipper is probably Australia's most specialized butterfly. It only occurs in one state, New South Wales, and only in subalpine areas in scattered parts of the Great Dividing Range, including the Blue Mountains. Its altitudinal span is a stingily restricting 800–1,300m (2,600–4,250ft), and here it is confined to woodland and heath, but only damp heath or near streams. The caterpillar only feeds on one species of native iris, *Patersonia sericea*; but even then it only likes one colour form of that iris, which has broader and bluer leaves than other forms. And to cap it all, the caterpillar only feeds in the afternoon and at dusk.

NON-BITING MIDGES
Diptera / Chironomidae

Male non-biting midges have superb, feathery antennae.

Midges can reduce the average human to a twitching, flailing, itchy mess. Summers spent near inland freshwater become an exercise in not being eaten. But not all midges bite. In fact, a huge number of midges are physically incapable. The non-biting midge family – Chironimidae – is enormous, with estimates of at least 10,000 species worldwide, making it one of the most ubiquitous insect groups on Earth. Male chironomids are marvellous-looking flies. They have fabulous, feathery antennae that they possibly use to pick up signals of females – possibly through pheromone or wing-beat frequency – across breeding territories. The non-biting midge is an extraordinarily adaptable insect, having settled in almost every aquatic habitat the world over, from tree holes to lakes. Their larvae develop in water, before emerging as winged adults. Their success has made them such a critical part of the global food web across the globe that the ecosystem might not function without them.

Mountain Beaver Flea

***Hystrichopsylla schefferi* | Siphonaptera / Hystrichopsyllidae**

This flea lives on the picturesque Pacific Coast.

The world's largest flea is a colossus among its kind. The female mountain beaver flea measures around 1cm (¼in) in length – around the size of a shieldbug, and it absolutely dwarfs other fleas. Once spotted, it is not easily missed, but this is an elusive flea that lives on an elusive host. Indeed, this flea has rarely been seen by humans. Mountain beavers live a secluded life deep in the forests of the North American Pacific Coast. They are not actually beavers, but a related rodent about the size of a musk rat, though they do display beaver-like behaviour in gnawing at tree trunks. The fleas live within the deep, dense fur, occasionally dropping off into nest material. The mountain beavers carry their oversized passengers, seemingly oblivious to the fact that their tenants are record breakers.

Sloth Moth

Bradypodicola hahneli **| Lepidoptera / Pyralidae**

Up to 100 adult sloth moths may reside in the fur of a single sloth.

The name says it all: sloth moths do indeed live on sloths. *Bradypodicola hahneli* lives in the hair of just one species, the pale-throated three-toed sloth (*Bradypus tridactylus*). Up to 100 individuals may piggyback on a single individual and can easily be seen running around the fur.

Nobody knows yet what the adults feed on, but there is no doubt what the larvae relish – sloth dung. The peculiar mammals are famous for defecating once a week, and while they do this, the moths fly down and lay eggs on the faeces. Once a sloth returns to the latrine, freshly emerged adult moths fly up and take up their unusual residence.

Dung Beetles
Coleoptera / Scarabaeidae

A world without sanitation doesn't bear thinking about; these guys lend a hand for free.

World Toilet Day highlights the need for clean sanitation for everyone, all over the planet. There is much poo in the world, and it has to go somewhere. For animals that don't have sewage infrastructure, this is dealt with in other ways. Since the first animals started to poo, a niche became available which has been filled by a host of invertebrates, bacteria and fungi that have evolved the ability to extract nutrients from dung, that cannot be absorbed by larger animals. Most well known are the dung beetles, which roll poo into almost perfect spheres. They collect the dung of large herbivores, pressing and forming it into balls and roll them back to their subterranean nurseries for their larvae to feast upon. All over the world, at any one time, millions of insects are consuming poo from every species (including, yes, humans); if they weren't, we'd all be literally up to our necks in poo.

Stalk-eyed Flies

***Diopsis* spp. | Diptera / Diopsidae**

Behold, the magnificent headspan of the male stalk-eyed fly.

Flies are amazing. They come in every conceivable shape and size, with all manner of evolutionary accessories and extras. But having your eyes mounted onto the ends of long stalks is a truly extraordinary and seemingly bonkers adaptation; nevertheless the stalk-eyed flies have managed it. There are a number of species from different fly families with this 'hypercephalization' – a broadened head with eyes on the ends of tubular stalks, rather like a hammerhead shark. It is only seen in males, who use the ocular appendages to attract females. They will gather in groups, lekking and having boxing matches with each other to claim mating rights; in turn the females will 'check out' the males with the most impressive heads.

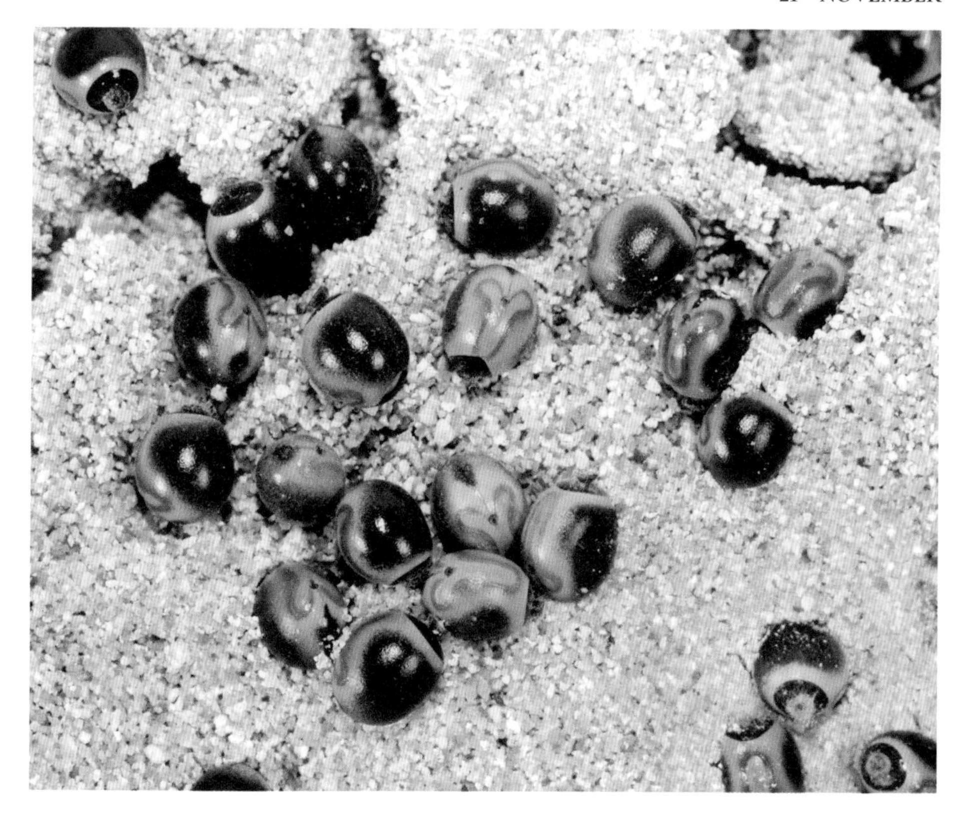

Stick Insect Eggs

Phasmatodea

These are the seed-like eggs of the phasmid *Tiracoidea biceps*.

Stick insects are justifiably famous for their amazing camouflage, their stick-like bodies and limbs so slender that they appear to be part of the vegetation. What many people don't realize is that this ability is not just confined to the adult stages, but to the very first stage – the egg.

The eggs vary between species and are intricately marked – a selection of them look like handicrafts from a craft shop. They look like seeds, and that is the intention – many have a fatty section at one end that imitates a similar edible structure in a seed. The idea is that ants are fooled into bringing them into the safety of the nest, where they are safe from predators, and once the ants have consumed the fatty part, they won't be disturbed. The phasmid nymph initially resembles an ant and leaves the nest unharmed.

BRUSH JEWEL BEETLE

Julodis cirrosa | **Coleoptera / Buprestidae**

Stiff, waxy hairs give this beetle a distinctly punk vibe.

The Buprestidae, or jewel beetles, are a diverse bunch. Many species in this most numerous beetle family have highly metallic and colourful elytra, but some choose to project their personalities in other ways. The brush jewel beetle, which lives on the hot, dry plains of southern Africa, is a delightful, walking toothbrush. Its otherwise subdued, black exoskeleton is bristling with stiff, waxy hairs (setae) that stand up in tufts of orange and yellow, presumably to advertise its (genuine or pretend) lack of palatability to predators. The result of this artifice is a punk rock vibe that would be the envy of 1977 Carnaby Street.

Wallace's Sphinx Moth

Xanthopan praedicta **| Lepidoptera / Sphingidae**

The proboscis of this moth is up to 30cm long.

We've all heard of celebrity endorsements – but how about celebrity predictions? One day in 1862, Charles Darwin, who was writing about orchid pollination, was sent a specimen of a remarkable flower from Madagascar with an extraordinarily long nectar tube, measuring 30cm (12in) in length. His first thought on seeing the star orchid (*Angraecum sesquipedale*) was 'Whatever could pollinate this?' Six years later, no less a naturalist than Alfred Russel Wallace (who had co-written the original theory of evolution paper with Darwin) declared that it would be a hawk-moth and stated: 'That such a moth exists in Madagascar may be safely predicted, and naturalists who visit that island should search for it with as much confidence as astronomers searched for the planet Neptune – and they will be equally successful.' The moth concerned was duly found and described in 1903.

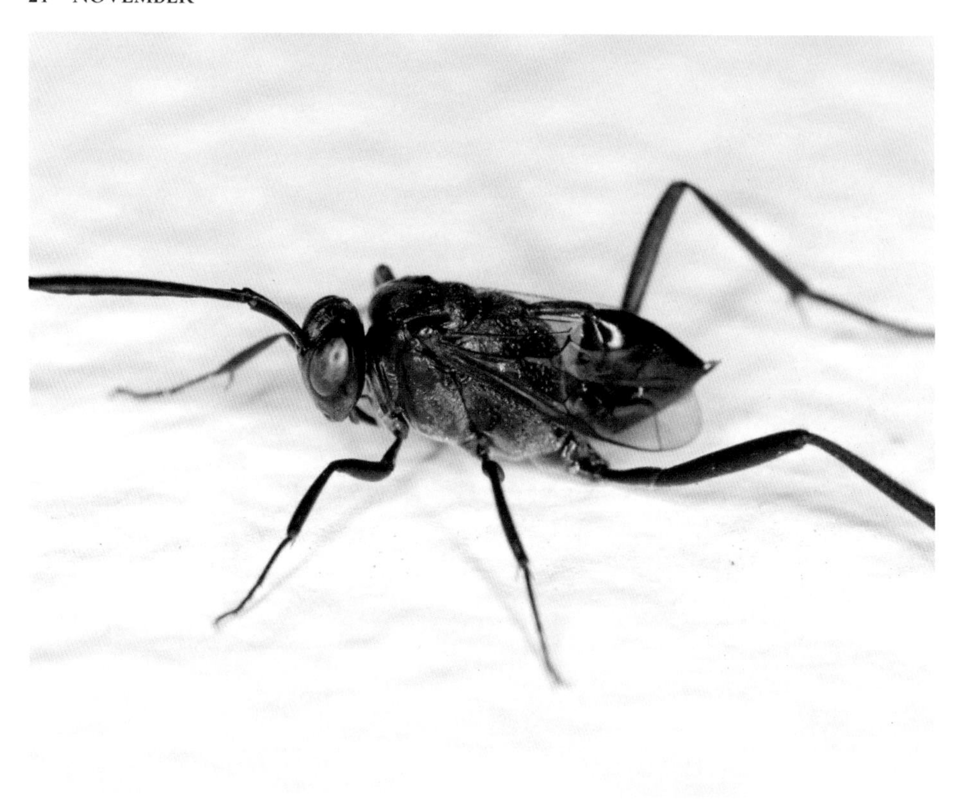

Blue-Eyed Ensign Wasp

Evania appendigaster **| Hymenoptera / Evaniidae**

Cockroaches can survive most things, but many meet their match in this parasitic wasp.

'What's the point of wasps?' This is a question that passes the lips of most people at some point or other. The question really being asked is: 'What purpose do wasps serve to humans?', because we humans are rather transactional and need to know what's in it for us. The answer to this question is manifold: pollination, ecosystem engineering, medicine. But still wasps are vilified, because their benefits are actually too large to see – they are intangible unless you can see the biggest-of-the-big picture. So how about this completely tangible benefit that surely nobody can ignore or misunderstand: wasps kill cockroaches. Yes they do – the Evaniidae family of ensign wasps are specialists at despatching roaches in the most efficient way possible, by targeting the egg sac, and my goodness, are they efficient. Just a single egg is laid on the egg sac (ootheca) of a live or dead adult cockroach; the wasp larva that emerges is capable of despatching the lot.

Fungus-growing Termite

Macrotermes natalensis **| Blattodea / Termitidae**

Many species of termites grow their own food underground.

Termites are among the most successful herbivorous insects in the world, and one of their favourite foodstuffs is fungi. But rather than just wandering out of the nest to find this precious tipple, they have evolved the capacity to grow their own inside their mounds. Over the course of deep time, estimated at 30 million years, the termites and a fungus called *Termitomyces* have become mutually dependent. The termites bring dead plant material from outside to culture the fungi, while the fungi provide much needed nitrogen in the form of nodules, which the insects eat. The workers collect spores of *Termitomyces* from the outside, place them in wood on the mound floor and even provide them with growth hormones. It's another example of how insects have been practising agriculture before humanity even evolved!

Great Crested Grasshopper

Tropidolophus formosus **| Orthoptera / Acrididae**

This grasshopper has a super-smart crest down the back of its thorax.

The great crested grasshopper deserves a special mention, just for that spectacular eponymous neckwear. Both sexes have this magnificently enlarged neck but can be told apart by the wing length which is long in males and shortened in females. Nothing transforms an otherwise innocuous grasshopper into a fearsome unit like an enormous crest, and this beast cuts a very stylish swathe on the southern Great Plains of the United States.

Warrior Wasps

***Synoeca* spp. | Hymenoptera / Vespidae**

Warrior wasps work together closely to protect their colony, and pull no punches when threatened.

Warrior wasps are social and come from the same family as paper wasps and hornets. Unlike conventional social wasp colonies, however, which have one queen, warrior wasp nests contain multiple queens, and so their lifestyle is more akin to the honey bees. When the colony becomes too large, a number of new queens leave and part of it splits off with them, swarming until a suitable nesting location is found. This behaviour has been observed in several closely related species and research does suggest a higher degree of brood success using this social plan. Warrior wasps are fiercely protective of their young; if threatened, they gather on the surface of the nest and collectively 'drum', to ward off predators. If this is ignored, they will attack and sting, but they do share another honey bee characteristic: the sting is barbed, and is pulled violently from the body on deployment, resulting in the unfortunate wasp's demise.

New Zealand Glow-worm

Arachnocampa luminosa **| Diptera / Keroplatidae**

The delicate, biolumininescent strands of glow-worm larvae hang from the ceiling of river caves in New Zealand.

While most insects use bioluminescence to attract partners, the New Zealand glow-worm uses it to catch food. *Arachnocampa luminosa* is an endemic fungus gnat and the only species found on Aotearoa (the country's Indigenous name). It lives in dark, riparian forest, in the overhanging canopy or in caves, where the reduced light levels are the perfect stage for it to set its trap. The glow-worm larva produces and suspends a sticky thread of globules from the cave ceiling or tree branch. These globules contain reactive enzymes that give off their own light, attracting small, photophilic insects, which then become trapped in the sticky garlands and are picked off by the hungry larvae. When many larvae construct their glow-traps in one place, the result is an incredible light show, which now attracts thousands of visitors each year to the river caves of the North Island.

Antbutterflies

Mechanitis, Melinaea **spp. | Lepidoptera / Nymphalidae**

Antbirds provide a highly unusual food source for some forest butterflies.

Tropical forests are complex, interdependent ecosystems, and some of the associations they throw up are amazing. Take the antbutterflies, three related clear-winged species from Costa Rica. They attend the swarms of predatory soldier ants that march across the forest floor eating everything small in their path. The butterflies often drop in among the hordes and seemingly put themselves at risk. What are they doing? Well, there is a cohort of birds that also follow army ants. They don't eat the ants themselves, but they feed around the periphery of the columns, catching insects fleeing for their lives. Several species of birds, some of them antbirds (Thamnophilidae), forage in no other way. Every day they attend the columns.

And the butterflies? They are there to feed from the antbirds' droppings. It's a predictable supply of scarce nutrients!

Giraffe-necked Weevil

Trachelophorus giraffa **| Coleoptera / Attelabidae**

Weevils are well known for having long noses, but this one has a super-long neck.

Having an extremely long, slender neck has its advantages. The giraffe can reach places other browsers cannot, exploiting a food niche to its advantage. Male giraffe-necked weevils have a similarly impressive expanse between body and head (females are somewhat more modestly proportioned), but it's not for food. The males with the longest necks are considered the best prospective partners, as they use their long necks to roll leaves up for females to lay their eggs in, providing a safe place for the larvae to develop and feed. These weevils are tiny – less than 1cm (¼in) long – but those marvellous necks can increase the males' size to a whopping 2.5cm (1in). They do have something in common with giraffes, though – in both species the males use their extensions as a weapon to batter each other; the winner gaining the opportunity to secure the attention of the females.

DECEMBER MOTH

Poecilocampa populi **| Lepidoptera / Lasiocampidae**

In the dark days of December in northern Europe, it's good to know there are some moths flying around.

If you're a moth enthusiast in Britain and northern Europe, you find yourself inordinately attached to those few moths that fly in late autumn and winter. In the cold, dark and relatively invertebrate-less days of December, it is truly comforting to know that there is a December moth, and it flies right from October deep into its eponymous month. It's a smart species, too, dark brown with a hint of purple, and with pale pink bars. It's the colour of a winter hot drink.

The caterpillars come in two colour forms, dark brown and variegated. The latter look lichen-coloured, perhaps for camouflage.

Science's Unlikely Hero

***Drosophila melanogaster* | Diptera / Drosophilidae**

The fruit fly was the first living organism sent into space.

A tiny fruit fly called *Drosophila melanogaster* has played a key role in advances in human biology and chemistry. So, what makes *D. melanogaster* the perfect lab buddy? Well, they are very easy to look after, and very small. They are sexually dimorphic (males and females look different), which is useful for biological sex-specific research. They reproduce quickly; each female lays hundreds of eggs over the course of a few days, providing a regular supply of flies, and several generations can be studied over a short period of time. Crucially, its DNA is 60 per cent identical to human DNA, and 75 per cent of genes that cause diseases in humans are also present in these flies, meaning that fruit fly and human have a remarkably similar make-up. It was the first insect to have its genome fully mapped and the first living organism that we sent into space, (beating Laika the dog by a decade). And it is there again now, orbiting the Earth in the International Space Station.

Hercules Beetle

Dynastes hercules **| Coleoptera / Scarabaeidae**

A sculpture of a male Hercules beetle outside the Bristol Aquarium, UK.

This improbable insect could surely come only from the most biodiverse and extravagant place on Earth – and so it does, living in the vast tropical forests of the Amazon Basin north to Central America. It is one of the largest and heaviest beetles in the world and, owing to the male's remarkable 'horns', is the longest. How it flies in the dark without hitting something can only be imagined.

Not surprisingly, the males use their two pincers, one from the head and the longer, curved one from the thorax, in combat. In captivity rivals try to ensnare and slam each other, but in the wild they might simply try to knock each other off branches or logs.

Monarch Butterfly

Danaus plexippus | **Lepidoptera / Nymphalidae**

The monarch butterfly undertakes what is undoubtedly the most famous insect migration in the world.

The world's most famous migration of an insect takes a pause about now. Most of the monarch butterflies from east of the Rocky Mountains have now arrived in Mexico's Central Highlands, where they gather in their millions, crowding on to the branches of oyamel firs (*Abies religiosa*) at an altitude of 3,200m (10,500ft). From a range of 2.6 million km (1.6 million miles) in the summer, they are now concentrated into a region measuring just 30 × 60km (18½ × 37¼ mi). They will remain here until March. Some have migrated all the way from southern Canada. The same butterflies will stir in the spring and migrate north again to the Gulf States, where they breed and die. Some will have covered 5,000km (3,100 mi).

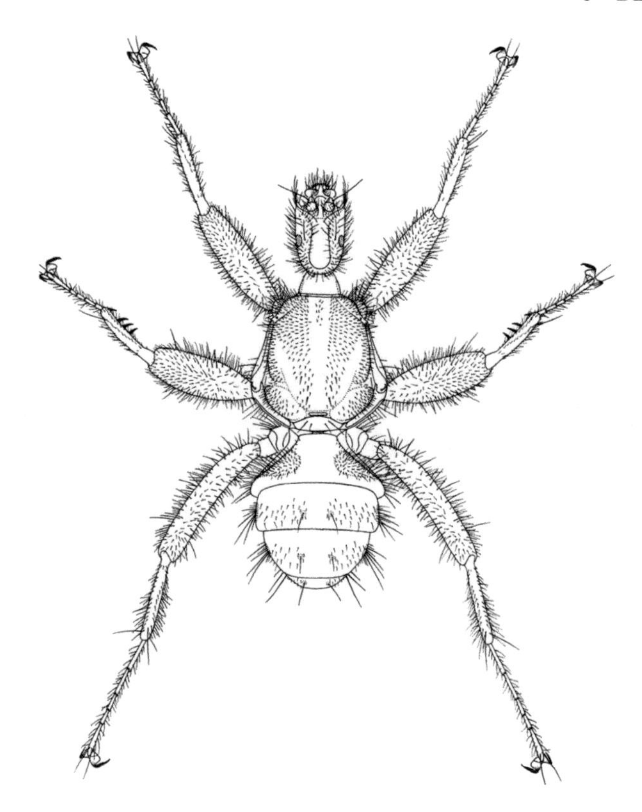

NEW ZEALAND BAT FLY

Mystacinobia zelandica **| Diptera / Mystacinobiidae**

New Zealand bat flies lives with bats, which could in theory predate them.

If ever there is a niche existence, how about being a New Zealand bat fly? There are only two species of land mammals in New Zealand, both bats, and the local bat fly is only interested in one, the lesser short-tailed bat (*Mystacina tuberculata*). It eats the colony's guano at the bottom of the tree-hole. These flies are downright strange. They are wingless. And blind. The flies groom each other, and their larvae. Even the larvae groom larvae!

Elderly male flies, which are beyond their reproductive age, do something that almost no other male insects do: they carry on living. The flies live in dangerously close proximity to insectivorous bats, but these males form a special protective caste that produces loud, high-frequency calls to tell the bats to keep off. Of all the quirks of this bat fly, having a use for old males is perhaps the most unusual of all.

Arctic Woolly Bear Moth

Gynaephora groenlandica | **Lepidoptera / Erebidae**

Sometimes a fur coat isn't enough – internal antifreeze is required too.

Do you feel the cold easily? Then you might want to take a leaf out of the Arctic woolly bear moth's book. It lives in the cold expanses of northern Canada and along Greenland's tundra. It has evolved an ingenious way to avoid freezing to death. Besides having a dense pile of insulating hair all over its body, the larva also contains cryoprotectants – chemical compounds that work like a biological antifreeze – which means that it can survive in temperatures as low as –70°C (–94°F) without being frozen solid. The larva will spend up to 90 per cent of its pre-pupal life cosied up in a dormant state, emerging on the scant amount of available warm days to inch its way up onto rocks, where it will bask in the sun and feed on whatever vegetation it can find in the brief Arctic summer.

Cathedral Termite

Nasutitermes triodiae **| Blattodea / Termitidae**

Some termite mounds in Australia are thought to be over 200 years old.

The famous cathedral mounds of Australia's Northern Territory are among the tallest insect-made structures in the world, with some of them reaching 8m (26ft) above the soil. Most workers are only 3mm (1/8in) tall, so the mounds are the equivalent of four times the height of the Burj Khalifa, the world's tallest building (830m/2,700ft), in Dubai. The mounds are prominent and long-lasting features of the countryside, and Aboriginal folklore suggests that some are at least 200 years old.

And that poses a question. Termite colonies have a long-lived queen and king, but how old can they get? It is thought that some live 70 years or more, the greatest age recorded for an insect.

THORN BUG

Umbonia crassicornis | **Hemiptera / Membracidae**

This cluster of thorn bugs could easily be mistaken for a stem of thorns.

On its own, this thorn bug merits a long second glance, but many clustered together really are a spectacle on another level. This inhabitant of southern Florida is a treehopper: a large family of short, stubby true bugs with a mind-boggling array of evolutionary adaptations. *Umbonia crassicornis* looks like a large thorn, which, when it sits still on vegetation, makes it look like part of the plant itself – a very useful way to evade predators. When large numbers of these treehoppers aggregate, a plain twig is transformed into a savage-looking limb, which may be of mutualistic benefit to the plant by making it look far less appetizing than it actually is.

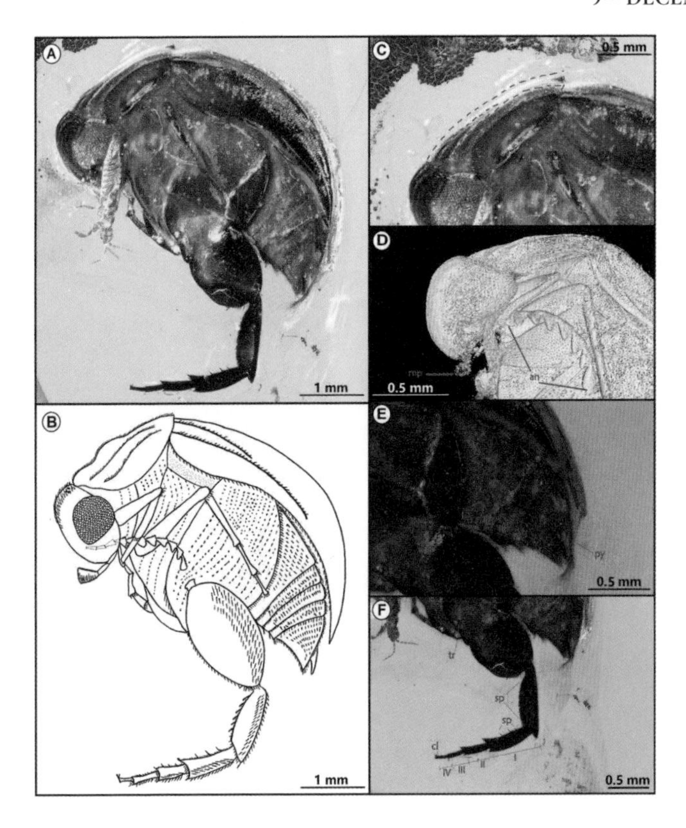

The First Pollinators

Angimordella burmitina **| Coleoptera / Mordellidae**

This beetle gives us an incredible glimpse into what the world looked like in the time of the dinosaurs.

As things currently stand, Earth's earliest recorded pollinating insect is a beetle that became trapped and encased in amber a long, long time ago. Discovered around 99 million years later in a mine in modern-day Myanmar, microscopic examination of the fossil revealed the prerequisite anatomical characteristics of modern pollinating beetles, and even found particles stuck to its exoskeleton that turned out to be the tiny pollen grains of prehistoric flowers. Named *Angimordella burmitina*, this beetle is part of the large family of tumbling flower beetles, the Mordellidae, which has evolved to encompass over 1,500 species worldwide. It is truly amazing to discover that, to this day almost 100 million years later, tumbling flower beetles continue their long tradition of pollinating flowers all over the world in a process that has barely changed since dinosaurs inhabited the planet.

Sawtooth Planthopper

Cathedra serrata | Hemiptera / Fulgoridae

Just when we think we've got a grasp of evolution and phylogenetics, something like this comes along to bamboozle us...

We humans like to name things, but evolution is a funny old thing. The duck-billed platypus defied the rules of classification, confounding scientists for some time. Many believed that the platypus was a hoax until it was eventually confirmed that it was a real, undoctored animal. The insect world is full of equally rule-bending species. For example, why on Earth the sawtooth planthopper would ever evolve like this is a mystery. In what looks like the product of a genetic experiment, this inhabitant of tropical South America is part bug, part moth, part sawfish, its head extended forward into a denticulated nakiri. The exact purpose of this chainsaw-like appendage may not be fully understood, but what it does affirm is just how lucky we are to live among such extraordinary diversity of life.

Mydas Flies

Gauromydas spp. | Diptera / Mydidae

The colossal mydas flies of the Amazon are the largest flies on Earth.

Gauromydas is the largest described genus of flies on Earth. They can measure up to 3cm (1¼in) in length, with a 10cm (4in) wingspan, dwarfing most other flies. These elusive creatures live in the Amazon rainforest, where the larvae live as secondary parasites in the nests of leafcutter ants, feasting not on the ant grubs, but the larvae of beetles, which themselves eat detritus from the floor of the nest. Despite their formidable size, *Gauromydas* do not interact with humans and pose no threat. These giant dipterans can look rather scary at first sight; they are black and shiny with orange-brown wings and a pointy face, resembling large hunting wasps (a cunning adaptation to protect themselves from predators) which, when they barrel through the trees towards you would cause an adrenaline spike in the boldest of us. Rest assured, the adults are harmless nectar feeders, and it is thought that females may not feed at all, saving their energy instead for mating and ovipositing.

Silk Moth

Bombyx mori | Lepidoptera / Bombycidae

Silk moths have been used for silk production for 7,000 years.

Silk is one of the world's most important insect products. It is made by the fifth instar of the larva of the silk moth as it prepares a pupal cocoon, and it takes 3,000 silkworms to produce 1kg (2¼lb) of the product. Silk production has been much criticized in recent times because most of the larvae are boiled alive inside their cocoons. Nevertheless, it's a truly ancient industry, known from China at least 7,000 years ago. A few other wild species are sometimes used, and moves are afoot to make the production more ethical.

Lantern Bug

Pyrops intricatus | **Hemiptera / Fulgoridae**

Nobody knows for certain what the long 'nose' is for – but it doesn't emit light, as was once thought.

How could anyone look at a picture of a lantern bug and not fall in love with insects? This wonderful species occurs in the rainforests of Borneo, where it lives a life of drinking sap from the stems of trees and exuding honeydew from its rear end. This particular species is associated with just one species of tree, known as the mata kucing (cat's eye) fruit tree (*Dimocarpus longan* ssp. *malesianus*).

The very odd foreparts, particularly the long, upward pointing hollow 'nose', are presumably an adaptation to help it get its food. When first discovered, the peculiar shape gave rise to the name 'lantern bug', because people thought that the snout emitted light. But this is a complete myth.

The Ant and the Grasshopper, by Charles H Bennett.

The Ant and the Grasshopper

This is one of Aesop's Fables, ideal for a cold winter's day. The story concerns two ways to spend the summer of plenty. The ant works hard all day long, working for its living, and storing up enough food to get through the winter. The grasshopper, on the other hand, spends the summer partying, dancing, hopping and singing, keeping no thought for the expected hardship ahead. Eventually, in the grip of winter, the grasshopper knocks at the ant's door, begging for food. But it is banished away, hungry. The moral of the story is clear to see.

Entomologists will be delighted that, in the original, the grasshopper was actually a cicada.

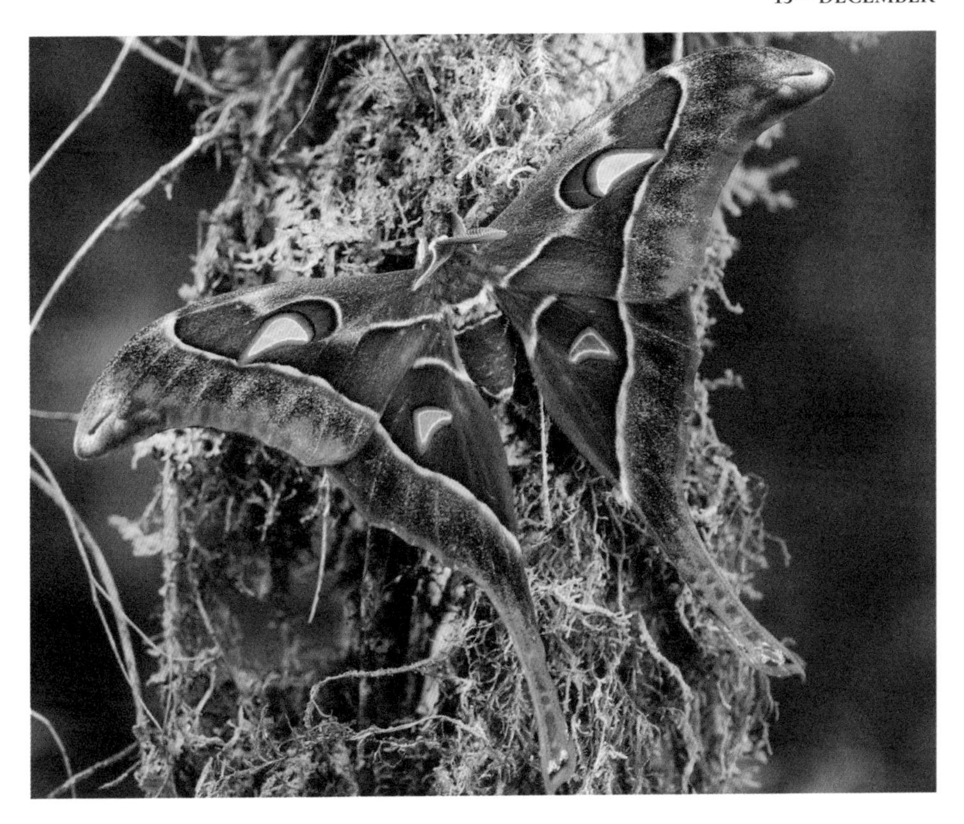

Hercules Moth

Coscinocera hercules **| Lepidoptera / Saturniidae**

This moth currently holds the record for the largest wing surface area of any insect.

The Hercules moth is not the largest, longest, or heaviest insect in the world, but it does hold the accolade for having the largest wing surface area of any insect currently described – an impressive 300 sq. cm (46½ sq in). It is endemic to New Guinea, and northern Australia.

The caterpillar is large and pale green, covered with statuesque spikes, and is an expert eating machine. It needs to consume vast amounts of food because, once it retreats into its chrysalis to pupate, it will never eat again. Adults do not have mouthparts and survive off the remaining fat reserves from their caterpillar days.

Fig Wasps
Hymenoptera / Pteromalidae

Male fig wasps aren't just aggressive, they are positively murderous.

Fig wasps are tiny wasps that live in fig flowers and play an important role in pollination. Often a particular fig species relies on one species of wasp to pollinate it. The females can fly from fig to fig, but the males are flightless and must wait for females to appear.

Things often get hot within the fig. Male wasps are exceptionally savage. If females are in short supply, they fight without hesitation. In one study, 50 per cent of the entire population of males were killed by their peers, one of the highest rates in the animal kingdom.

Austral Ellipsidion Cockroach

***Ellipsidion australe* | Hemiptera / Blattodea**

This cockroach looks as though it has been exquisitely hand decorated.

Cockroaches. Almost universally reviled by humans for their links to anthropogenic activity, exponential reproduction and apparent indestructability (remember the Twinky-eating 'roach in *Wall-E*?). While it is true that many non-native species have made themselves extremely unwelcome by virtue of simply being successful, there are also numerous species of cockroach that never interact with humans. They are also incredibly beautiful. The Austral ellipsidion is native to eastern Australia. The adult is a small (less than 2cm/¾in), delicate-looking beast with intricate, latticed wings like fine lacework, bettered only by the nymph, which, with its rich black and terracotta bodywork punctuated with white dots, looks like a stunning example of Indigenous Australian storytelling. This cockroach feeds on honeydew, mould and pollen, and is found in forests and urban spaces, but poses no threat to humans, except maybe to knock us out with its sheer beauty.

LEICHHARDT'S GRASSHOPPER

Petasida ephippigera **| Orthoptera / Pyrgomorphidae**

The locals associate this grasshopper with the thunder of tropical storms.

This has to be a candidate for the world's most beautiful grasshopper, and one look should be enough to convince any predator not to eat it – with bright colours like this, the message is 'Leave me alone, I'm poisonous'. The Leichhardt's grasshopper secretes a noxious fluid too, just in case.

It lives in the Top End of Australia's Northern Territory, where it is rare and declining. One of its few strongholds is Kakadu National Park, where the local Kundjeyhmi call it *alyurr* and regard it as a child of the lightning man Namarrkon, who responds to the insects' calls with the storms of the tropical summer.

Leichhardt's grasshopper was discovered by Westerners in 1843 during the third voyage of HMS Beagle, the same ship that carried a young Charles Darwin on its previous outing.

Dalcerid Moth

Acraga coa | **Lepidoptera / Dalceridae**

Orange is the new black for this funky moth.

Those of you familiar with that arts and crafts staple, pipe-cleaners, will already be nodding your heads earnestly and saying, 'Yes, yes, I know exactly where you're going with this.' This delightful little moth, from the rainforests of Central and South America, could easily be the work of a pre-school art project. Some orange pipe-cleaners, craft paper and a healthy dose of imagination, *et voilà*, we have *Acraga coa*, a fabulous, tangerine moth with delicate, feather-boa legs and a fuzzy body enveloped in a sleek, sunset cape. And the fun doesn't end there – the larva of this moth is known as the 'jewel caterpillar' and it could not be a starker contrast. If the adult looks like an art project, the caterpillar is glossy and translucent, injected with gummy bear colours that have come straight out of the workshop of the finest glass workshops in Murano.

Ventilator Termite

Odontotermes obesus | Blattodea / Termitidae

Engineers have studied termite mounds to learn about ventilation.

The large termite mounds that dot the ground in many areas of the tropics, looking like little jagged peaks, are not only remarkable feats of construction by tiny, blind insects; they are also engineering marvels. Below ground, where the queen resides and where the termites grow fungi for sustenance (see page 327), the temperature is kept remarkably constant at 30°C (86°F). It's done by an internal convection system. Near the roof of the mound, there is a maze of narrow conduits, while in the middle of the mound is a large chimney. By day, the air in the roof warms up quickly, while air in the chimney sinks. By night, however, the air in the roof cools, plunges down and replaces the air in the chimney, which escapes through the conduits. This brilliant arrangement has been incorporated into human buildings.

Freeze-tolerant Fly

***Heleomyza borealis* | Diptera / Heleomyzidae**

In the Arctic tundra, there is a fly that can survive temperatures well below zero.

Heleomyza borealis isn't just hardy, it can survive temperatures not many other animals can withstand. According to research, it is a rare example of a 'freeze-tolerant polar arthropod'. It times its larval stage (the development phase of an insect most resistant to cold conditions) to coincide with the Arctic winter, when it has reached a minimum weight and has accumulated the necessary quantity of lipids to enter a hypometabolic state before the temperature plummets from just above zero, to less than –20°C (68°F). The larvae remain in suspended animation until they detect a specific pattern of temperature rises, upon which they pupate and emerge as adults in the warmest weeks of the Arctic year.

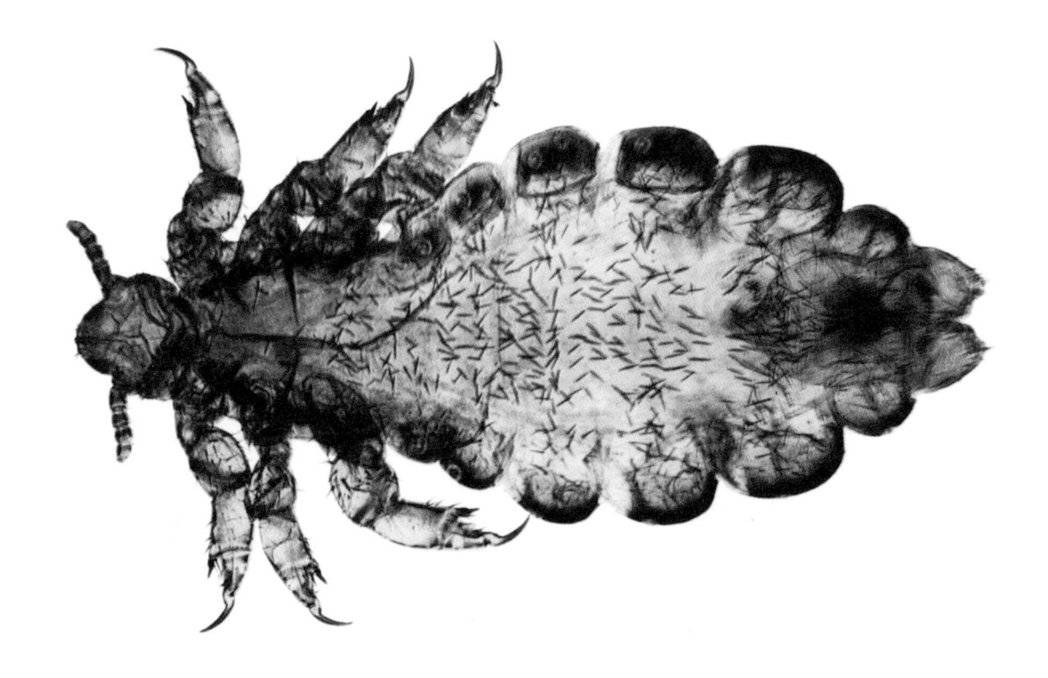

Human Louse

***Pediculus humanus* | Psocodea / Pediculidae**

Head lice and body lice are two forms of the same species.

You might itch as you read this. The human louse is a host-specific insect that feeds on the blood of humans, taking a meal five times a day. It occurs in two forms. The head louse lives on the scalp and is transmitted by close contact, combs and headgear, since it cannot fly and spends its whole life cycle on one host. The eggs are glued on to strands of hair. The body louse is another form of the same species; it really should be called the clothes louse, because it bases itself in fibres of clothing, crawling on to the body only when it needs to feed. While head lice are annoying, body lice can be deeply unpleasant, and can also be vectors for disease. Geneticists have worked out that the two forms diverged about 170,000 years ago and it is reasonable to conclude from this that humans must have adopted clothing at approximately the same time.

Snowflakes in the Rainforest

Hemiptera / Flatidae

These fabulous planthoppers twitch and leap around when disturbed.

It's pretty unusual to see snow in the tropics, but the sight of planthopper nymphs leaping around might just make you doubt your own eyes. These pre-adult hoppers from the Flatidae family of bugs, which spend their time drinking sap from plant stems, have the ability to secrete a waxy substance from their rear ends. This secretion holds its form like semi-hardened Silly String to create an outfit that wouldn't look out of place at Rio Carnival. The waxy filaments are thought to act as a defence against predators, possibly by extending their radius of sensitivity. If touched they will twitch and bounce and, if further agitated, leap clean off the stem while their assailant is left behind with nothing more than a mouthful of wax.

A portrait of Charles Dickens.

The Cricket on the Hearth

'Heyday!' said John, in his slow way. '[The cricket is] merrier than ever, to-night, I think.'
'And it's sure to bring us good fortune, John!' [replied Dot] 'It always has done so. To have a Cricket on the Hearth, is the luckiest thing in all the world!'
'The first time I heard its cheerful little note, John, was on that night when you brought me home. Its chirp was such a welcome to me! It seemed so full of promise and encouragement. It seemed to say, you would be kind and gentle with me.'
'I love it for the many times I have heard it. Sometimes, in the twilight, when I have felt a little solitary and down-hearted, John, its Chirp, Chirp, Chirp upon the hearth, has seemed to tell me of another little voice, so sweet, so very dear to me, before whose coming sound my trouble vanished like a dream. It has cheered me up again and filled me with new trust and confidence. I was thinking of these things to-night, dear, when I sat expecting you; and I love the Cricket for their sake!'

Adapted from *THE CRICKET ON THE HEARTH* BY CHARLES DICKENS (1845)

Christmas Beetles

***Anoplognathus* spp. | Coleoptera / Scarabaeidae**

Nobody can deny that seeing this beetle would be an awesome Christmas present.

Santa Claus is traditionally viewed as the star guest in Australian homes at Christmas time, but if you're really lucky you'll also receive a visit from a Christmas beetle. These charming scarab beetles are so named because, many years ago, they would come out in large numbers throughout Australia, most frequently along the length of the east coast of Victoria and Queensland, right down to Tasmania. They come in a variety of 'designs' and colours – some are so shiny they wouldn't look out of place on the Christmas tree. Christmas beetle larvae live in soil, where they eat grass roots and build up enough energy to pupate; the emergent adults ascend to the tree canopy to feed on eucalyptus leaves. Unfortunately, the Christmas beetle's occurrence has decreased significantly as the urban sprawl of modern cities reduces access to suitable habitat. Help these glorious beetles by letting them share your lawn so they can brighten up your holidays!

Midges

Forcipomyia spp. | Diptera / Ceratopogonidae

This midge has a mind-boggling wing beat.

Some facts about the natural world are almost too astonishing to comprehend. One of these concerns a thoroughly obscure biting midge called *Forcipomyia*. Laboratory measurements have shown that this insect can beat its wings 1,046 times *a second*.

Yes, a second. Think about that for a second. And if you do, *Forcipomyia* will have flapped its wings 1,046 times.

Common Earwig

Forficula auricularia | **Dermaptera / Forficulidae**

This is the view the baby earwigs get.

Out in the garden, under a rock, a sweet domestic scene plays out, hidden from our eyes. A female earwig has dug a nest chamber and is tending her clutch of 50 or so eggs. She spends most of the time dormant in the sealed burrow, but at times will turn the eggs over and lick them, evidently to apply fungal disinfectant (neglected eggs quickly grow mould). When spring comes and the eggs are about to hatch, she spreads them out to ease the process, and once the nymphs emerge, she gathers them all under her body, fetching any that stray with her mouthparts, just as a cat does. As they grow, she feeds them assiduously, finding food on the surface and offering it by regurgitation. Eventually, the nymphs leave, thoroughly pampered, the products of the tenderest of care known from any insect.

Eos and Tithonus

Above: The story of Eos and Tithonus is testament to the importance of semantics.

Opposite top: The Antarctic midge lives on the continent all year round.

Opposite bottom: Not even the snow puts off the indomitable winter gnat.

Be careful what you wish for. It is something that a certain mythological deity should have heeded when she approached her father for a favour. Eos, the Greek goddess of the dawn, and daughter of Zeus, fell in love with the Trojan prince, Tithonus. However, their chance of happy-ever-after was impeded by immortality – specifically his lack of it. Eos approached her father, Zeus, to beg him to give Tithonus the eternal life she dearly wished for him. Zeus duly dispensed the gift, and all was well until Tithonus's eyesight and faculties went a little squiffy. Eos realized, horrified, that she had asked for eternal life, but not everlasting youth. Her lover aged and withered before her eyes, and eventually she did what any loving, bereft partner would do in these circumstances: she transformed him into a grasshopper (or a cicada, depending on which source you read). Grasshoppers are believed to be immortal in mythology, so Eos was forever bound to hear the constant refrain of Tithonus's chirping across the meadows.

29TH DECEMBER

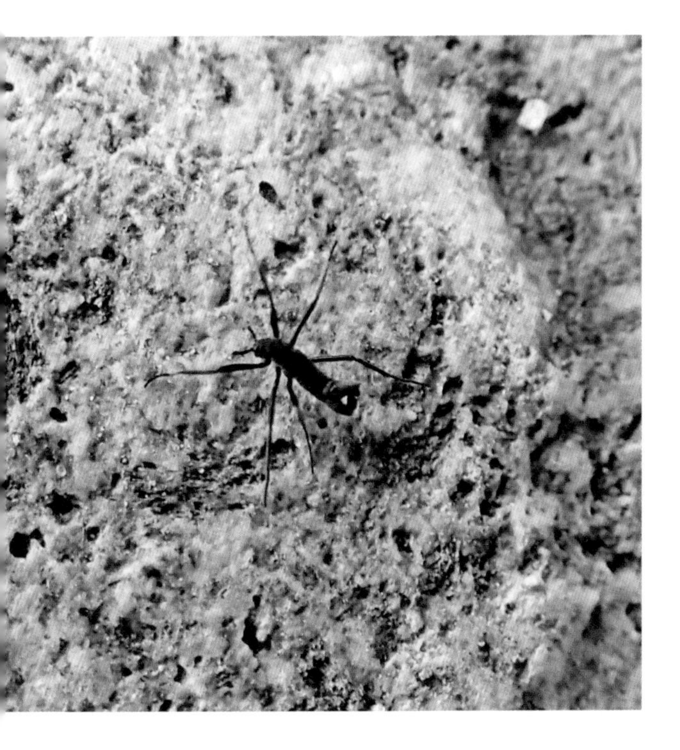

ANTARCTIC MIDGE

Belgica antarctica

Diptera / Chironomidae

It says something about the conditions found in Antarctica that this flightless midge is the largest living terrestrial animal that lives year-round on the continent. Everything else flies or swims away for part of the year. It lives in the soil, the adults only emerging in the summer. The eggs are covered with antifreeze jelly and the overwintering larvae have antifreeze compounds in the blood. Strangely, tourists don't usually mention this unique insect among their 'bucket list' highlights.

30TH DECEMBER

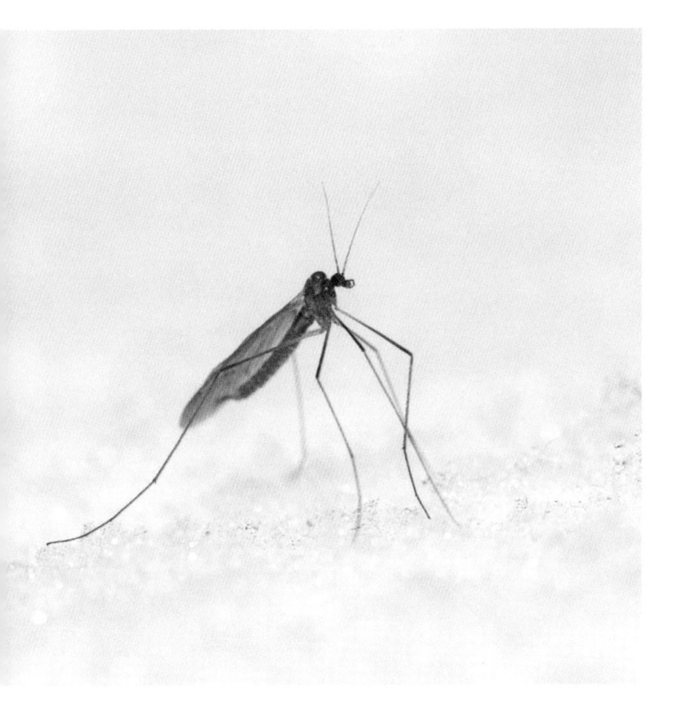

WINTER GNAT

Trichocera annulata

Diptera / Trichoceridae

Do we dare whisper it – that the midwinter is made to feel just a little less bleak by a gnat? It's a cold winter's day in Europe or North America, but in a sunny corner of the garden, or a sheltered field edge, where the rays warm the frosty air, there is a small gathering of tiny flying insects, dancing up and down.

31ST DECEMBER

BLACK BARK MANTID

Paraoxypilus tasmaniensis **| Mantodea / Nanomantidae**

The enlarged forelegs of this mantid make it look like its wearing boxing gloves.

As if spending your life as the potential prey of a praying mantis among vegetation isn't stressful enough, imagine how nerve-shattering it must be to live with the ultra-cryptic boxer bark mantids, which include the black bark mantid. These small (approx. 2cm/¾in) insects live on tree trunks and among leaf litter in Australia. They have a rugged, mottled brown body, which gives them supreme camouflage, rendering them virtually imperceptible against the gnarly bark. Males are long-winged, allowing them to fast-track themselves around the trees in search of females, which are flightless and instead put all their energy into egg-laying. The characteristic 'praying' forelimbs of this group of mantids are unusually thickened, giving them a pugilistic appearance, which has earned them the amusing, though slightly misleading, nickname 'boxer mantids'. They do not actually box, although with their unparalleled patience and lightning reflexes, they would probably be pretty awesome at it...

Index

Acknowledgements

I would like to extend my thanks to Tina Persaud for commissioning this work in the first place, and to the talented team at Batsford for getting it into shape – especially Hattie Grylls, Rebecca Armstrong and Tally de Orellana. I also owe a debt to my family for their wonderful support during this challenging but fantastic project. – Dominic Couzens

I would like to thank Rebecca and the team at Batsford for their incredible dedication and love for this book. Thanks to my family and friends for their emotional support and snack deliveries during its creation, particularly Damion, who is the right to my left. – Gail Ashton

Picture Credits

Every effort has been made to contact the copyright holders. If you have any information about the images in this collection, please contact the publisher.

© 1, 77 Mike Walker / Alamy; 2, 3, 332 Rolf Nussbaumer Photography / Alamy; 4, 204, 233 mauritius images GmbH / Alamy; 5 Razvan Cornel Constantin / Alamy; 7 Mark Horton / Alamy; 8, 177, 245, 312 Richard Becker / Alamy; 11, 192 Arcangelo Manoni / Alamy; 12, 79 David Forster / Alamy; 13, 28 Prisma by Dukas Presseagentur GmbH / Alamy; 14, 179 Dorling Kindersley ltd / Alamy; 15 Tom Stack / Alamy; 16, 312 robertharding / Alamy; 17 The History Collection / Alamy; 18, 67 Mary Evans / Natural History Museum; 19 Custom Life Science Images / Alamy; 20 Doug Wechsler / Nature Picture Library; 21 MYN / Gil Wizen / Nature Picture Library; 21, 41 Survivalphotos / Alamy; 22, 23 Keren Su/China Span / Alamy; 24 Bence Mate / Nature Picture Library; 25 Joana Wiesbaum de Paiva Boléo, licensed under the Creative Commons Attribution-Share Alike 3.0 Unported license (https://creativecommons.org/licenses/by-sa/3.0/deed.en); 26 Actep Burstov / Alamy; 27 CC BY CBG Photography Group 2016, Piotr Naskrecki / Nature Picture Library; 29 Brian Scantlebury / Alamy; 30, 81, 83, 84, 93, 133, 152, 222, 242, 272, 284, 318, 347 Nature Picture Library / Alamy; 31, 217 Bryan Reynolds / Alamy; 32, 296 Penta Springs Limited / Alamy; 33 Marco Uliana / Alamy; 34 Roland Bouvier / Alamy; 35 Genevieve Vallee / Alamy; 36, 247 Florilegius / Alamy; 37 Marcus Harrison - food / Alamy; 38 AGAMI Photo Agency / Alamy; 39 Mitsuaki Iwago / Nature Picture Library; 40 MURILO GUALDA / Alamy; 42, 62, 99, 100, 107, 114, 117, 144, 165, 185, 191, 193, 215, 260, 303 Gail Ashton; 43 Dan Leeth / Alamy; 44, 101, 125, 208, 210, 218 NPL - DeA Picture Library / Bridgeman Images; 45 James Cutting; 46 Purix Verlag Volker Christen / Bridgeman Images; 47 ulrich missbach / Alamy; 48, 51 World History Archive / Alamy; 49 Wattlebird / Alamy; 50 Constantinescu Costin / Alamy; 52, 134, 161, 171, 234, 240, 275 Minden Pictures / Alamy; 53 bilwissedition Ltd. & Co. KG / Alamy; 54 Amazon-Images / Alamy; 55 Samantha Muir / Alamy; 56 Phil Savoie / Nature Picture Library; 57 Clarence Holmes Wildlife / Alamy; 58, 119, 361 Hakan Soderholm / Alamy; 59, 79, 138 Nigel Cattlin / Alamy; 60 Florilegius / Getty Images; 61 Australian Associated Press / Alamy; 63, 86 Granger / Bridgeman Images; 64 Drägüs, licensed under the Creative Comons Attribution-Share Alike 3.0 Unported (https://creativecommons.org/licenses/by-sa/3.0/deed.en); 65, 288, 352 Premaphotos / Alamy; 66 Photononstop / Alamy; 68, 198, 326 Tomasz Klejdysz / Alamy; 69, 181, 257, 300, 324, 325 The Natural History Museum / Alamy; 70 fish1715 / Shutterstock; 71, 143, 158, 268 Papilio / Alamy; 73, 337 Desmond W. Helmore, licensed under the Creative Commons Attribution 4.0 International license (https://creativecommons.org/licenses/by/4.0/deed.en); 74, Leena Robinson / Alamy; 75 Cayambe, licensed under the Creative Commons Attribution-Share Alike 4.0 International license (https://creativecommons.org/licenses/by-sa/4.0/deed.en); 76 James Peake / Alamy; 78 Valter Jacinto / Getty Images; 80 Ian West / Alamy; 82 Carim Nahaboo; 85 Nature Collection / Alamy; 87, 305, 334 SDym Photography / Alamy; 88 EDU Vision / Alamy; 88, 140, 313 Library Book Collection / Alamy; 89 History and Art Collection / Alamy; 90 Tim Gainey / Alamy; 91, 199, 264, 315, Larry Doherty / Alamy; 92 Cavan Images / Alamy; 94 CPA Media Pte Ltd / Alamy; 95 Florilegius / Mary Evans; 96 Jason Ondreicka / Alamy; 97, 238, 243, 298, 308, 311 BIOSPHOTO / Alamy; 98 The Robin Symington Collection/Mary Evans Picture Library; 103 YongXin Zhang / Alamy; 104 Cephas Picture Library / Alamy; 105 De Luan / Alamy; 108 Bird MS, Bilton DT, Perissinotto R, licensed under the Creative Commons Attribution 4.0 International license (https://creative-commons.org/licenses/by/4.0/deed.no); 109 Mark Marathon, licensed under the Creative Commons Attribution-Share Alike 3.0 Unported license (https://